Introduction to Architectural Technology
Second edition

Pete Silver
Will McLean

with contributions by
Dason Whitsett

D1511275

Introduction to Architectural Technology
Second edition

Pete Silver
Will McLean

with contributions by
Dason Whitsett

Laurence King Publishing

LAURENCE KING

First published in 2008 by
Laurence King Publishing Ltd
361–373 City Road
London EC1V 1LR

Second edition published in 2013
by Laurence King Publishing Ltd

Tel: +44 20 7841 6900
Fax: +44 20 7841 6910
email: enquiries@laurenceking.com
www.laurenceking.com

A catalogue record of this book is available from the
British Library.

ISBN 978 1 78067 295 3

Design: Hamish Muir
Picture Research: Annalaura Palma

Printed and bound in China.

Dedicated to the memory of
James Madge

Right
National Aquatic Center, Beijing, China
(detail), designed by CSCEC
International Design and Architects
Peddle Thorp Walker in conjunction
with Arup Engineers and Vector Foiltec.

Contents

Foreword

In the course of my practice as a structural engineer and as a visiting teacher at a variety of architecture and engineering schools, I have encountered much discussion, debate, information, and misinformation about the role of "technology" in architecture. Many use the subject to discuss the different ways in which buildings are put together, while others draw on the digital realm – where the materials are processes, networks, and flows that have been enabled by the pace of change in computers as a design tool.

What is clear to me is that the best systems prosper when both of these approaches are exploited in design development – as intellectual activities that integrate production, assembly, and creation to bring together quantative and qualitative satisfaction in the built environment.

At one extreme very little practical illustration is published that shows, for example, the role of basic structural systems or basic ways of using materials. On the other hand it is common to see students produce amazing diagrams of flow or stress analysis that then magically produce "architecture" at the other extreme.

What is needed is the middle ground, and in this respect, this book provides easy access for the beginning student of architecture. Learning through a selection of contemporary architectural projects introduces students to basic systems and appropriate use of both new and old materials, of both techniques and tools.

What the authors have carefully catalogued is the fundamental information that the architecture student must know, and to be aware of the need to work with experts in various disciplines and to learn from past precedents. This will give them a better chance of designing with originality and novelty without becoming an engineer or computer scientist. They will learn to produce architecture by facilitating dialogue between experts, manufacturers, politicians, and communities.

At a time when we are not short of information, what is needed is the skill to edit the information to suit the purpose; something which this book achieves with great success.

Hanif Kara, F.I. StructE, FRIBA (hon)
Adams Kara Taylor
November 2007

Left
The USA Pavilion, Montreal, Canada (detail), designed by R. Buckminster Fuller and Shoji Sadao.

Introduction

"Aerodynamics is for people who can't design engines." Enzo Ferrari

Both structurally and environmentally, the architect designs the way in which buildings are organized. Along with the art – the social, cultural, philosophical, and political factors that influence design – it is the tools and techniques of the engineer – technology – that gives form and performance to the built environment.

Critical to an understanding of architecture is the relationship between design and technology. The aim of this textbook is to introduce the scope of that relationship to those who may be considering a career in architecture, as well as to enable students of architecture to integrate their design thinking with appropriate structural and environmental solutions.

The book sets out to explain the relationships between physical phenomena, materials, building elements, and structural types, using simple classification systems and real examples. Photographic images are used to familiarize the user with common construction technology, while historical examples are employed to chart significant moments in the history of architectural engineering.

The book introduces structural and environmental engineering to architects and, while it does not include mathematical calculation, it does make reference to current computer techniques that assist designers in predicting the structural and environmental behavior of buildings. At the same time, it makes liberal reference to historical precedents since it is important for the user to understand that the way in which the success of technology is measured is directly related to its cultural context. Some would say that the Gothic cathedral is the pinnacle of architectural engineering given the tools, materials, and techniques of the day. Currently, the world is far more conscious of the clean and efficient use of the planet's resources, and the success of technology is measured thus.

Technology by nature is concerned with invention, and it is only ever possible to take a "snapshot" of technological achievements at any one moment in time. From materials science through to construction site procedures, invention, and progress in architectural technology occur constantly; it is hoped that a compendium of technology such as this book will not only aid students to understand its scope and historical context, but will also inspire them to invent new solutions for the future.

Second Edition

This second edition includes new information on the fast growing field of Building Information Modeling (BIM), as well as a section on the ways in which computer control may be used to enable buildings to adapt and respond to environmental changes. New diagrams have also been added to many of the case studies, and typical details have been included in the section on weathering. Under Building Services, there are new sections on integrating active and passive systems and on sanitation.

11 Structure of the book

Architectural technologies are classified under two major headings: **Structure and Form**, and **Climate and Shelter**. While the very first primitive shelters, such as mud huts and igloos, employed the same materials and building components to both create structural form and withstand the elements, even a simple tent such as the North American tepee employed one type of material and component (timber posts) to generate the form and another (animal skins) to maintain its internal climate.

Structure and Form
This section is subdivided into three main topics: **Structural Physics**, **Structural Elements**, and **Structural Logic**. This is a linear progression from the fundamental principles of structural form – materials science, materials technology, and structural integrity – through descriptions of the common elements used in the construction of buildings, such as walls, columns, floors, and roofs, to how these elements are combined to form different building types, from simple "post and beam" structures through Gothic vaults to aluminum monocoque shells and tensile nets.

Climate and Shelter
The subdivisions in this section deal with two main topics: **Human Comfort** and **Building Performance**. Before examining how buildings are constructed in relation to their local environment, it is important to establish the thresholds within which the human body is able to sustain life. The first part of the section looks at the principles of thermal comfort and its provider, the sun.

The second part considers ways in which the built environment can be designed to maintain human comfort. As well as exploring how the design and fabric of a building – the climatic envelope – can both control and interact with its environment, passively and actively, it considers human comfort in the context of sound and light.

The book includes three additional sections:

Computational Tools and Techniques
The ways in which architects design are now much influenced by computers. Quite apart from their use in drafting and illustrating, they may assist designers to predict both the structural and environmental behavior of their buildings – through modeling and analysis – as well as supplying databases for materials, components and systems; indeed they can store, and create a virtual model of, all of the information needed for a construction project. This section explores the use of computers as analytical and organizational tools.

Case Studies
A range of case studies is used to explore the origins of construction types and illustrates the historic relationship between design and technology in architecture and civil engineering. The examples are listed chronologically and are described in relation to the structural and environmental taxonomies described in the first two sections.

Building Codes
Throughout the world, the construction industry is controlled by rules and regulations that are there to ensure the health and safety of both construction industry operatives and end-users. Building codes cover all aspects of construction and, increasingly, are concerned with regulating towards low-energy, sustainable solutions.

Contents Page
This is designed as a branching system so the user can contextualize any topic within the book. Every spread has a topic heading that can be cross-referenced with the contents tree.

Captions
The main body of the text is supplemented by illustrations, both diagrammatic and photographic. Captions are used to supply additional information and should be read as a continuation of the main text.

Taxonomy Charts and Icons
There are two major taxonomy charts used in Structure and Form, one for Structural Elements and one for Structural Logic. The organization of the charts is directly related to the page spreads that follow, and the individual diagrams are then employed as iconic references. These icons are also employed to cross-reference case studies with the main body of the text.

Further Reading
Related publications are listed under each of the major headings. It is also recommended that users reference the source information, both publications and web sites.

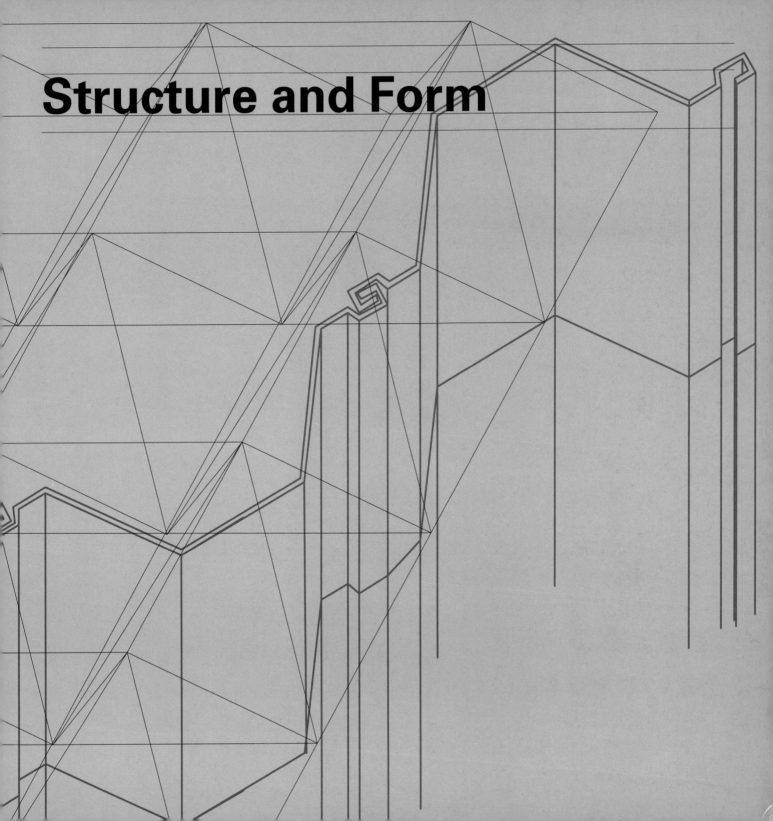

Structure and Form

¹⁵ Structural Physics

Covering the fundamental principles of structural form

Materials Science

In order to understand how buildings work as structures, it is necessary to appreciate the natural laws that govern structural form. This involves an understanding of the forces that act upon materials and the ways in which the materials will react to those forces according to their fundamental properties. Materials Science covers: forces – how, through gravity, materials are subject to stress and strain; properties of materials – how materials can be analyzed and manipulated for strength and stiffness; and reactions – how components behave when subjected to forces through bending and shear.

Materials Technology

While materials science is the study of the interplay between the properties, structure, and changes in solids, materials technology is the utilization of this knowledge in the production and fabrication of materials and structural components. Most fabrication processes involve a series of primary ways of acting on materials – cutting, sawing, bending, welding, etc. – with the precise nature of the process determined by the properties of the material to which it is applied. In general, the ability to move raw materials and finished products around factories, building sites, and indeed around the world, places finite limitations on the size, shape and weight of prefabricated building components. Materials technology explores a range of common building materials: stone, timber, steel, reinforced concrete, glass, fabric, fiber-reinforced plastic, and sheet materials.

Structural Integrity

Knowing the behavior of the materials and components that go to form structures enables the architect to organize them in such a way that the whole structure will be rigid and stable under load. Structural Integrity examines the types of load to which buildings are subjected – live and dead loads, lateral loads, bracing, and torsion – and how structures may be constructed to resist them through triangulation and bracing. Overall stability, as in the rooting of a tree, is then considered in the context of the center of gravity and the cantilever (or branching) principle.

Left
The leaning tower of Pisa, Italy, 1173.
The tower is 55.8 m high and it leans at
an angle of 5.5 degrees.

¹⁶ Materials Science / Forces: Stress, Strain

Newton's Second Law of Motion

Structural engineering is concerned with calculating the ability of structures and materials to withstand forces. When you talk about the weight of an object, you are actually describing the force that the object exerts as a result of gravity acting on its mass. Gravity is responsible for attracting all objects towards the center of the earth. Over 300 years ago, Isaac Newton discovered that this gravitational "pull" (indeed, all forces) produce an acceleration of the system or object that they are acting upon, hence:

FORCE = MASS (lb) x
ACCELERATION (ft per sec., squared)

Force is measured in Newtons:
1 Newton = 1 kg x 1 m/sec^2
(2½ lb x 3 ft 3 in/sec^2)
Gravity is calculated as:
g = 9.81 m/sec^2 (g = 32 ft 2 in/sec^2)

Structural Form

All objects have an underlying structure that maintains the form of the object when it is subjected to forces. The forces act upon the object, and the object reacts according to the properties of the materials from which it is made and the way in which they have been shaped and assembled. Stress is the term used to describe the forces acting on a material or structure; strain is the term used to describe the material or structure's response to those forces.

Stress

Forces can act on materials in one of two fundamental ways: they can push, which is a compressive force, or they can stretch or pull, which is a tensile force. These applied forces are described in physics as stress, which is measured as the force per unit area across any given cross-section of the material:

f (STRESS) = P (FORCE) / A (AREA)

Strain

Materials use their internal structures to resist forces. Every material reacts to stress by distributing it in such a way that there is an equal balance of internal forces. The result is strain – a change in the form of the structure, measured as the fractional extension perpendicular to the cross-section. Hence, compression may be described as stress that acts to shorten an object and tension as stress that acts to lengthen it.

1 Compression
Bulgarian weightlifter Glagoi Blagoev's body is put into compression by the weights held aloft (approximately 430 lb or over twice his own body weight).

2 Compression
When force is applied to the top of the structure indicated by the arrow, the sides of the structure are put into tension (T), forcing the cross section to increase.

3 Tension
A tug of war puts the rope into tension over its length, forcing the sides of the rope into compression and thus reducing the cross-section of the rope. The ratio of cross-sectional contraction and linear extension is described as a Poisson ratio.

4 Tension
When most materials are put into tension, their cross-sectional area is put into compression (C) thus reducing the cross section.

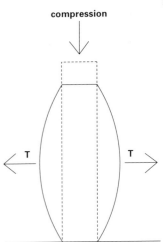

compression

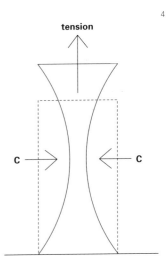

tension

18 Materials Science / Properties of Materials: Strength, Stiffness and Flexibility

Properties

Every material has its own internal structure that predisposes it to resist forces to varying degrees and to react to them in different ways. It is this behavior – this reaction to forces – that gives a material its inherent structural properties. Some materials, for example, will tear easily but can be pulled (like paper), and some can be pulled but not compressed (like string). Some, like concrete, are strong in compression but weak in tension and others, like steel, are strong in both tension and compression.

Plastic-Elastic Behavior

As it resists forces a material will react by deforming its internal structure. With plastic materials the internal structures readily deform and adapt to a new shape when stressed. A plastic material distorts easily but does not break (an example is lead). With elastic materials the internal structures remain the same but are warped or stretched; they always remember their original shape and want to return to it. An elastic material returns to its original length (or shape) when the stress is removed (an example is bamboo).*

Strength

While a fishing rod is not strong, it is very elastic – it will bend a great deal before it snaps. It is the ability of an elastic material to resist bending (elastic deformation) that will determine its strength. All materials have an elastic limit – if you stretch them far enough, they will fail. They will either warp or buckle, or they will break – fracture or tear. A strong material is one with a high breaking stress, a weak material one with a low breaking stress.**

Stiffness and Flexibility

A stiff material needs a large force (stress) to produce a small extension (strain) – it is difficult to change its shape. A flexible material only needs a small stress to produce a large extension – it is not difficult to change its shape.

* Materials with low plasticity are prone to breaking or fracturing under stress – are brittle – and are likely to become more so as temperatures decrease. Highly plastic materials deform without breaking or fracturing – are ductile – and are likely to become more so as temperatures increase. Heating metals increases their plasticity.

** Standard structural steel will fail at around 30 ksi, while wood will reach its ultimate stress at around 1.0 ksi or less depending on species and grade.

1 The strength and stiffness/flexibility of a material, as determined by its resistance to (elastic) deformation, can be measured as the ratio of stress to strain. When plotted on a graph, the (stiffness) values are known as the Young's Modulus. Young's Modulus is used to measure increases in a material's length when it is put under tension or to predict the buckling or yield point. Generally, as density increases so does Young's Modulus but there are exceptions to the rule.

2 The chart shows the relationship between categories of materials in respect to the Young's Modulus and the density of the material. Published by Granta Design, this type of chart was pioneered by Professor Mike Ashby, Cambridge University Department of Engineering. These charts manage to condense a huge range of analytical data into a graphically legible format within which the comparative properties and groupings of materials can clearly be visualized and assessed.

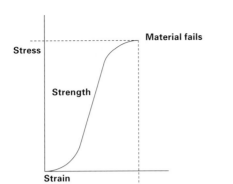

Stress

Material fails

Strength

Strain

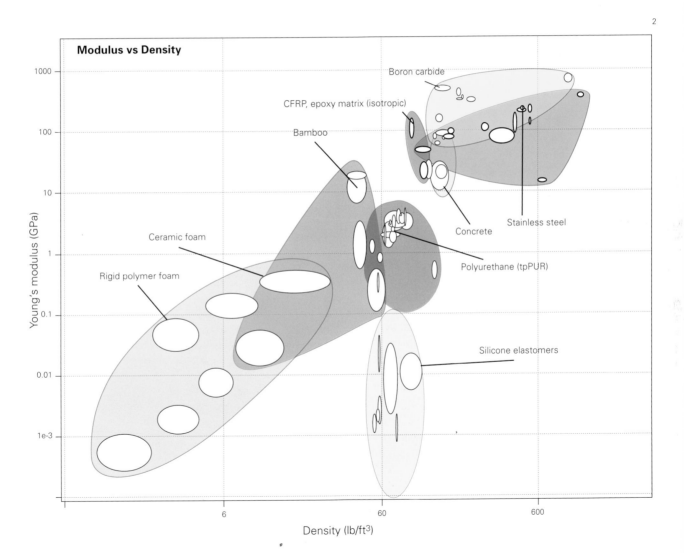

Modulus vs Density

Young's modulus (GPa)

1000

100

10

1

0.1

0.01

1e-3

Boron carbide

CFRP, epoxy matrix (isotropic)

Bamboo

Ceramic foam

Rigid polymer foam

Concrete

Stainless steel

Polyurethane (tpPUR)

Silicone elastomers

Density (lb/ft³)

6

60

600

20 Materials Science / Reactions: Bending, Shear

Bending

Bending is the result of a material deforming itself under stress. In diagram **1**, the beam at the top is subjected to an external, compressive force that in turn creates internal strain through bending (exaggerated in the diagram): the top side becomes compressed – shortened – and the underside becomes tensioned – lengthened. The lower diagram describes bending in a truss, showing the reactions of the individual components: red are in tension, blue in compression.

In diagram **2**, bending has been exaggerated by the insertion of cuts and wedges in the beam. Once loaded, the wedges in the upper side will be compressed and close up, while the cuts in the lower side will be pulled apart to form wedges. Excessive bending will cause failure.

Shape and Form

When materials are formed into structural elements, the distribution of stresses is modified by how they are shaped and formed – the deeper a beam, the further apart are these internal forces and the more even the distribution of stresses. Diagram **3** illustrates the effect of cross-sectional shape on bending.

Shear

Unlike tension and compression, where surfaces move towards or away from one another, shear refers to deformation in which parallel surfaces slide past one another, as shown in diagram **4**. Shear stress is a stress state where the stress is parallel or tangential to a face of the material, as opposed to normal stress when the stress is perpendicular to the face. In structural and mechanical engineering the shear strength of a component is important for designing the dimensions and materials to be used for its manufacture.

With an I-beam for example, the flange resists bending and the web resists shear. See Beam: Cross-Sections, page 44.

1 Effects of bending in a beam. The arrows on the top edge of the diagrammatic beam indicate the direction of forces under deflection. Similarly the top chord and the vertical members of the diagrammatic truss are put into compression under deflection (shown in blue), whereas the bottom chord and diagonal members of the truss are put into tension.

2 Effects of bending in a beam. A "notched" beam illustrates the expansion-contraction or compression and tension experienced by a beam under loading.

3 Effects of cross-sectional shape on bending. The orientation of structural members and their cross-sectional shape vastly affect their performance under loading. The pair of diagrams on the left illustrate how the orientation of a simple structural member can affect its loadbearing capacity. The increased depth that the orientation in the lower diagram shows vastly improves its structural performance. The drawings on the right show how an increase in cross-sectional profile stiffens a vertical member under load.

4 Shear forces act tangentially or parallel to a material face, rather than perpendicularly.

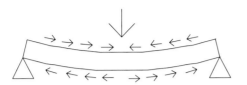

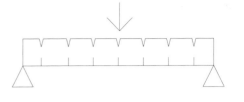

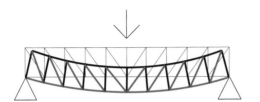

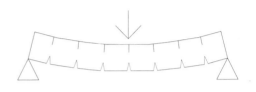

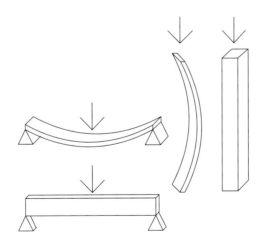

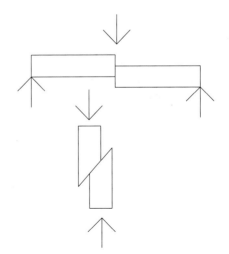

22 Structural Integrity / Loads: Live and Dead Loads, Lateral Loads, Bracing, Torsion

Live and Dead Loads

To create structural form, building components are sized, shaped, and arranged in such a way that the structure as a whole is able to resist the forces acting upon it. At this scale, a distinction can be made between the forces that arise through the mass of the structure itself – a constant force known as a static or dead load – and external forces that move or change, known as the dynamic or live loads. In buildings, live loads consist of furniture, machinery, occupants, and snow and ice, as well as, crucially, the lateral impact from wind.

Lateral Loads

A chair must be strong enough to resist the vertical load of its occupant. However, as the occupant shifts their load – to lean, wobble, or sway – the chair must remain rigid. This is because it is designed to resist these non-vertical loads through a combination of the stiffness of its components and the rigidity of the whole structure. This rigidity is, in effect, an equalizing of the vertical and horizontal forces throughout the structure by the use of diagonals – i.e. through triangulation.

Bracing

Take a simple, square frame. If you add a force to one side of it, it will deform (shear). If, however, you were to put a stick from one corner to the other, the square would remain stable. If you pushed on the other side of the square, the stick would still work but only if it was connected at both ends, as it is transferring the load by being pulled. In the first case, the stick is used as a compressive element; in the second, it is in tension and could be replaced with string or wire.

Torsion

Structural components may also have to be braced so as to prevent them twisting. Though sometimes referred to as a force in itself, torsion or twisting is in fact the result of opposing (tensile or compressive) forces acting in a rotary fashion.

1 Bracing in a steel-framed structure.
2 Principle of bracing. A square frame is an inherently unstable structure because of its lack of bracing. An applied force as indicated by the arrows will transfer the structure into a skewed diamond shape unless either a rigid (compression element) or a linear (tension) element is introduced diagonally across the frame to triangulate or "brace" the structure.
3 The key to bracing is triangulation, and even in a simple table there are small, "invisible" triangles bracing the structure.
4 "Pinwheeled" structure: this is an idealized structure where each of the legs or columns is widened to brace the structure and are then rotated so that the structure is able to resist forces from any direction.
5 Principle of torsion or twist, which can be resisted by the cross-sectional profile.

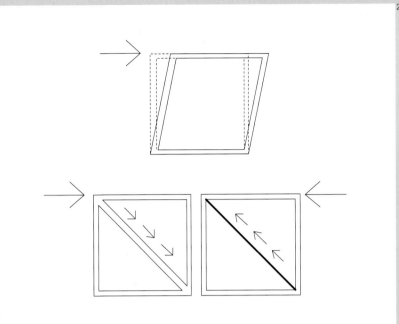

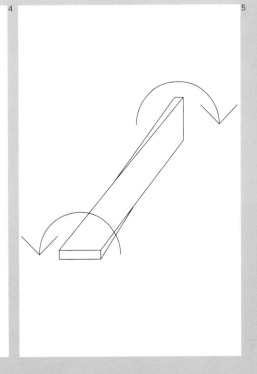

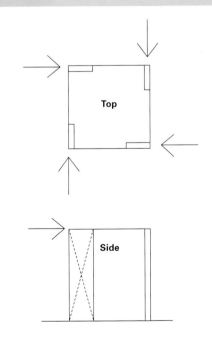

Top

Side

24 Structural Integrity / Stability: Anchorage, Height, Center of Gravity, Cantilever Principle

Anchorage

Buildings have to be braced in order to resist wind loads and, in more extreme circumstances, a sudden impact or vibration such as from a wave or earth tremor. Not only must the mass of a structure not deform under lateral loads, it must also not move as a whole. Buildings are anchored to the ground with foundations that also provide a bearing in unstable or soft ground.

Wind also produces internal pressure and suction on external surfaces. Roofs in particular need to be securely anchored to the main structural elements.

Height

Tall structures have to withstand not only high-speed gusts of wind, but also the vibrations that wind can induce (similar to those of a tuning fork). Tall buildings are designed to resist these forces by having a rigid, central core with floor plates that are carried from perimeter columns and connected back to the core. The core is also used to contain elevators, staircases and service ducts. Tall structures may also be stabilized by employing diagonal struts wrapped around the outside of the building in a criss-cross fashion; when used without a central core, these are known as tube structures.

A dense honeycomb of internal walls will also improve structural stability.

Center of Gravity

Try leaning forward from the hips. At some point your center of gravity goes "outside of you," and one leg moves forward to form the triangle that keeps you from toppling over – keeps you stable. Carry on bending, and you will reach the point when the only way to maintain your center of gravity is to extend your other leg behind you. This is a process known as "cantilevering".

Cantilever Principle

A cantilever is a description of an element that projects laterally from the vertical. It relies on counterbalance for its stability and on triangulation to resist the bending movements and shear forces of the lever arms.

1, 2 Reinforced-concrete cores provide stability for framed buildings.
3 The designers of the Forth Railway Bridge, Scotland used their own bodies to demonstrate how the span of the bridge uses the cantilever principle. The bodies of the two men at ground level are acting as columns (in compression), and their arms are being pulled (in tension). The sticks they hold are in compression and are transferring the load back to the "columns".
4 The Forth Railway Bridge with girder spans of 521 m, completed in 1890. Fife, Scotland.
5 Cranes displaying the cantilever principle.
6 Cantilevered brackets are employed to support a roof canopy at the Royal Ascot Racecourse, Windsor, UK by HOK Sport.

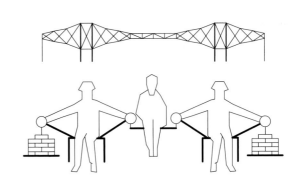

1

2

3

4

5

6

26 Materials Technology / Stone

Stone is a low-tensile material, i.e. it is strong in compression but weak in tension. Traditionally used to form loadbearing walls, columns and arches, it can be also be carved into lintels or beams, but since the material has little tensile strength and is dense and massive these spans tend to be of limited size.

The oldest of the building materials known to man, stone is still one of the few that is transformed into a product at source. It is excavated both from underground quarries accessed from shafts and by the opencast method where it is cleaved from the rock face using mechanical excavators. Different layers of rock (beds) produce different sizes, qualities, and types of stone, and this will determine the way in which it is acted upon. Stone can be sawn, cropped, split or chiselled into shape, although this material often has a grain which must be considered when maximizing its structural properties.

Stone blocks direct from the quarry are either cropped to size using a guillotine or sawn into shape using a range of mechanical saws. A multi-bladed frame saw will carve a block of stone into a set of slabs in roughly six hours. A diamond-bladed circular saw enables precision cutting as the blade follows a laser line, can operate in various axes, and is also used to mill or route precise geometric incisions from the surface of the stone. These tools can be instructed from computer-aided-design (CAD) files.

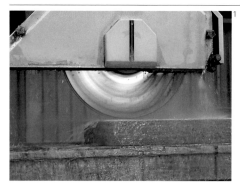

1 Diamond-bladed circular saw cutting a stone block.
2 Limestone quarry, Dorset, UK.

²⁷ Materials Technology / Wood

Wood in construction is most broadly categorized as rough lumber or finished lumber and as hardwood or softwood. Rough lumber is used for structural purposes while finished is exposed in the final product.

Hardwood and Softwood
Hardwood comes from deciduous trees and softwood from coniferous ones. Softwoods are used in most wood structural systems, light framing being the most common. The more rarely used heavy timber framing is usually also from softwood species.

Both hardwoods and softwoods are used for finished lumber. Lower quality finished lumber is often used when the surface may be patched and painted, while the higher quality material is used in applications where the wood will receive a clear finish.

To be used in construction, wood must be dried either in open air or in a drying kiln. Even after this process, the wood often contains enough moisture that it will continue to dry on the construction site. As wood dries, it shrinks.

Engineered Wood
Engineered wood uses wood fiber re-combined using glue, making a stronger product than solid wood. The most familiar engineered wood product is plywood, where thin veneers of wood are glued together in layers. Oriented strand board (OSB) frequently replaces plywood today as it instead uses chips of wood pressed and glued into a sheet. Engineered wood has also found applications in finished products, especially flooring.

Preservatives
When it is exposed, too much or too little moisture will cause wood to rot and become prone to fungal and insect attack. Except in the case of a few rot-resistant species, exposed wood must be painted, coated, or pressure-treated. Pressure treatments impregnate the wood with chemical preservatives and must be used where the wood is in contact with the ground or with concrete.

1 Lumber from the logging mill: rough sawn.

Materials Technology / Steel

Steel is a product of coal, iron ore and limestone, and is manufactured in a number of stages. The coal is rid of its impurities – transformed into coke – and mixed with iron ore and limestone in a blast furnace to produce liquid iron. This is known as pig iron and when solid it is strong but very brittle. To convert this into steel (which is strong but flexible) the proportion of carbon in the mix must be adjusted by blowing hot air or pure oxygen into the molten iron. Steel used for construction purposes is standard carbon steel, which has up to 0.3 per cent carbon and is known as mild steel. The liquid steel is then cast into billets that are the stock from which structural shapes are made.

Hot-Rolled Steel

Most of the steel used in the construction industry has been extruded through a profiling machine. In the hot-rolling process, the steel billets are re-heated as they are forced through a series of rollers to produce sections such as wide flanges, channels, angles, and bars. Steel sections can be curved by rolling. Sections can be connected by drilling and bolting (a process that can permit tolerance – for adjustment – or disassembly) or, for a permanent fusion, by welding. It is possible to create a fully welded joint that will fuse two pieces of steel so that they perform as one.

Most hot-rolled steel sections are specified according to their sectional designation, normal size, and weight per foot. For example, a W 8 x 28 is a wide flange approximately 8 in high that weighs 28 lb/ft. Hot-rolled sections are generally used for columns and beams.

Cold-Rolled Steel

Structural steel sections can also be formed by "rollforming" flat steel coil in a machine. Such machines are capable of producing a range of steel profiles out of ⅟₁₆-⅛ in thick steel coils. Cold-rolled sections are generally used for repetitive members that carry lighter loads

Corrosion

Corrosion is the deterioration of the essential properties of a material due to its exposure to the environment. It normally refers to metals reacting electrochemically with water and oxygen, as with the oxidization of iron atoms to produce rust. Iron or steel can be prevented from rusting in a number of ways: through surface treatments such as plating, painting, or the application of enamel or reactive coatings, or by galvanization – the process of coating iron or steel with a thin layer of zinc that protects the steel from rusting. Steel is either passed through molten zinc (hot-dip galvanizing) which oxidizes to form a surface layer, or zinc is deposited on the surface by electroplating (electrogalvanizing).

1 A range of hot-rolled steel sections.
2 Steel staircase fabricated from steel angles and flat "plate" strips.

29 Materials Technology / Reinforced Concrete

Formwork

Formwork is the name given to the mold from which concrete is cast. Formwork is usually built with dimensional lumber and plywood, but many prefabricated formwork systems are also available. Releasing agents are applied to the surface of the formwork before the slurry is poured. As it is placed, the concrete must be compacted by vibration in order to remove air bubbles.

Reinforcing

Concrete can be cast either on site – in situ – or in a factory – precast. In either case, although concrete is very strong in compression, the material must be reinforced with steel bars for it to withstand tensile stresses. The size and distribution of the stresses within the concrete are modified by the steel bars. The bars are formed into a cage which sits in the lower part of the beam where the tensile stresses are at their greatest.

Prestressed Concrete

For spans of over 26 ft 3 in it is advisable for the reinforcing bars to be tensioned. The principle behind prestressing is similar to that of lifting a row of bricks horizontally by applying pressure to the bricks at the end of the row. When sufficient pressure is applied, compressive stresses are induced throughout the entire row, and the whole row can be lifted and carried horizontally. With concrete, this can be done either by pretensioning or post-tensioning the steel reinforcement.

In pretensioning, the steel is stretched before the concrete is poured. Once the concrete reaches the required strength, the stretching forces are released and, as the steel reacts to regain its original length, the tensile stresses are translated into a compressive stress in the concrete. In post-tensioning, the steel is stretched after the concrete hardens. The concrete is cast around ducts or tubes, and once the concrete has hardened to the required strength, steel tendons are inserted into the tubes, stretched against the edges of the concrete, and anchored off externally, placing the concrete into compression.

Self-Compacting Concrete

Self-compacting concrete is a fluid concrete that will flow freely into molds and around steel reinforcement and will fully compact without the need for additional vibration. It contains additives that are known as plasticizers. Similar to washing-up liquids, plasticizers allow less water to be added to the concrete mix, making it stronger and ensuring an even distribution of the binder and aggregates. They also leave a glass-smooth surface both when the concrete is poured horizontally and when it is placed in formwork.

Concrete is made from a binding material combined with aggregates and mixed together with water to create moldable slurry. The binder consists of cement, granulated blast furnace slag or pulverized fuel ash, and aggregates consists of natural materials such as gravel, sand, and stone chippings.

The usual proportions are:
6 fine aggregate (e.g. sand)
1 coarse aggregate (e.g. stone)
1 binder (e.g. Portland cement)

1 Wet concrete (slurry) in a mixing machine.
2 Precast concrete component: casting colored concrete in a mold.

Float Glass

Float glass is a sheet of glass made by floating molten glass on a bed of molten tin. This method gives the glass uniform thickness and very flat surfaces. Modern window glass is float glass.

Float glass is made by melting raw materials consisting of sand, limestone, soda ash, dolomite, iron oxide and salt cake. These blend together to form a large pool of molten glass. Standard float glass comes in thicknesses ranging from ⅛ to 1 in, and in panes of up to a maximum size of 10 ft 6 in by 19 ft 6 in.

Glass used for structural purposes has been toughened in one or more of the following ways:

Annealing

By controlling the rate at which molten glass cools, the stresses and strains that normally occur due to uneven shrinkage during cooling can be removed from the glass.

Tempering/Toughening

Glass is re-heated and the surfaces are then rapidly cooled with jets of air. The inside core of the glass continues to cool and contract, forcing the surfaces into compression and the core into tension. When the glass breaks the core releases tensile energy, resulting in the formation of small glass particles.

Laminating

Layers of glass are bonded together with a resin interlayer to form a composite panel. If an outer layer breaks it is held in place by the substrate.

For safety reasons, structural glass beams typically consist of three layers laminated together. Beams are connected to columns using bolts and/or silicon glue. Structural glass is also used for ribs or "fins" that are fixed perpendicular to panes in order to resist wind load on glass façades.

1 Float glass being moved around a factory.

31 Materials Technology / Fabrics

Polymerization is the last in a series of chemical processes needed to create the molecular chains that form polyvinyl chloride (PVC). PVC is a thermoplastic and is produced in the form of a white powder that is blended with other ingredients to form a range of synthetic products. Additives include heat stabilizers and lubricants as well as those that determine mechanical properties such as flexibility.

Membrane fabrics are made from PVC-coated polyester. These synthetic fabrics are strong in tension and resistant to shear and there is a range of around 20 standard colors. They are light, flameproof and have a maximum thickness of about $\frac{1}{16}$ in. Ultraviolet light and weathering will cause deterioration in the fabrics over a period of between 10 and 20 years depending on the climate. Membrane fabrics can be used for roof canopies (see Tensile Surfaces, page 68) or, when pressurized with air, can be formed into structural elements such as air beams.

Air is not the only gas that can be used for pressurization. When a gastight fabric is inflated with hydrogen or helium it will form a buoyant (lighter than air) structure such as an airship.

The process of fabricating membrane structures is akin to that of tailoring. The fabric is delivered in rolls anything up to 7 ft 10 in wide, and panels can be cut by automated cutting machines according to computer-aided-design (CAD) profiles. These computer numerically controlled (CNC) cutters have a variety of interchangeable heads (consisting of a hole punch and cutting wheel) while the 69 ft benches along which they travel are designed to vacuum the fabric onto the cutting surface for precision cutting.

1 The design of pressurized or tensile membrane structures involves pattern-cutting techniques similar to those used in tailoring.
2 These polyester "air beams" have been stitched and welded together to form a continuous surface.

32 Materials Technology / Fiber-Reinforced Plastic

As with any pouring or molding process, the act of casting shapes out of fluid materials has the advantage that complex, curved surfaces may be achieved. Large-scale structural components are now commonly manufactured out of resins that are reinforced with fiberglass. The fiberglass is built up in layers inside the liquid resin – a process known as laminating – and products are often further reinforced with layers of foam, Kevlar or carbon fiber for extra stiffening.

The fiberglass itself is a material woven from extremely fine fibers of glass. It is commonly used in the manufacture of insulation and textiles, and is also used as a reinforcing agent for many plastic products, the result being a composite material called glass-reinforced plastic (GRP), also known as glassfiber reinforced epoxy (GRE). Fiberglass is usually supplied in the form of matting, known as chopped strand mat.

A resin is a plastic material that can change its state from liquid to solid under certain conditions, e.g. through adding a catalyst or hardener, or due to a temperature change. GRP uses epoxy or polyester resins, which are high performance resins that are strong, waterproof and resistant to environmental degradation. As opposed to resins which are cured by a catalyst, epoxy resins are a two-part adhesive in which the resin is mixed with a hardener.

Molding GRP
The mold is cast from a master pattern known as the plug and becomes the negative shape ("die") from which the final product is manufactured. Molds are either open or closed according to the type of product and the cost involved.

The process of casting begins with the application of a gelcoat, a barrier coating that can produce a finished surface with high gloss, color, and surface integrity retention. After the gelcoat, several layers of fiberglass matting and resin are applied as needed for the strength of the product.

Pultrusion
This process is similar to extrusion used with other materials, but pulls rather than pushes the material through a die to create a member with a constant cross section. In fiberglass pultrusions, continuous glass fibers in resin are pulled to create profiles that are used for construction applications. Some of the more common applications include window frames, grates, and structural shapes.

1 Fiberglass – chopped strand mat.
2 Fiber-reinforced plastic (FRP) products for the construction industry.
1 Pultruded fiberglass products.

33 Materials Technology / Sheet Materials

Traditionally cut with automated bench saws, sheet materials are now commonly cut to a very high tolerance using laser beams, water jets or plasma cutters. Computer numerically controlled (CNC) laser cutting is a process in which a shape is cut from sheet material (wood, plastics, aluminum, mild steel) using an intense laser beam that cuts by melting the material in the beam path. CNC lasers are housed in x/y/z-axis gantries that can be programmed to cut from computer-aided-design (CAD) drawings like a large plotter. A laser can cut through a 3 ft 3 in long, ⅜ in thick, steel panel in around 30 seconds.

Lasers are accurate to within 50 microns and the parts remain flat.

Once malleable materials (like sheet metal) have been cut into shape, they can be formed into complex, three-dimensional shapes by multi-axis folding machines. An 80-ton press brake can fold sheet metal up to ⅜ in thick. Sheets are folded by being compressed over interchangeable tool heads, and fabricators make use of extensive tooling "libraries" to construct multi-tool set-ups for each product. CNC machines can be programmed to carry out folding procedures over up to 8 axes, and, according to the thickness of the material, programs help calculate the tolerances to be made for the natural curvatures formed along the edges.

Once folded into shape, components are often assembled and finished by hand. They may be welded, riveted, or bolted together and can then be subjected to a variety of surface treatments such as plating, engraving, or silk-screening.

1 Laser-cut ¾ in sheet steel.
2 Laser cutting profiles from an acrylic sheet.

Bearing Elements

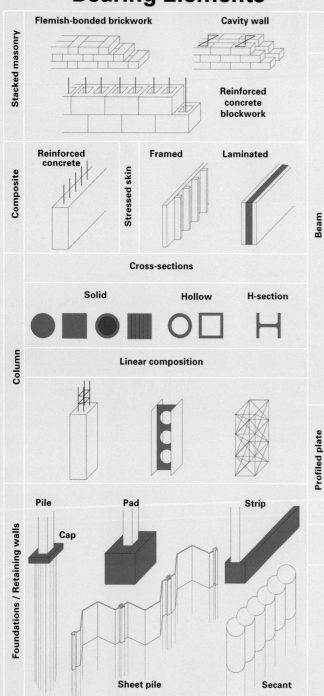

Stacked masonry

Flemish-bonded brickwork

Cavity wall

Reinforced concrete blockwork

Composite

Reinforced concrete

Stressed skin

Framed

Laminated

Column

Cross-sections

Solid

Hollow

H-section

Linear composition

Foundations / Retaining walls

Pile

Cap

Pad

Strip

Sheet pile

Secant

Spanning Elements

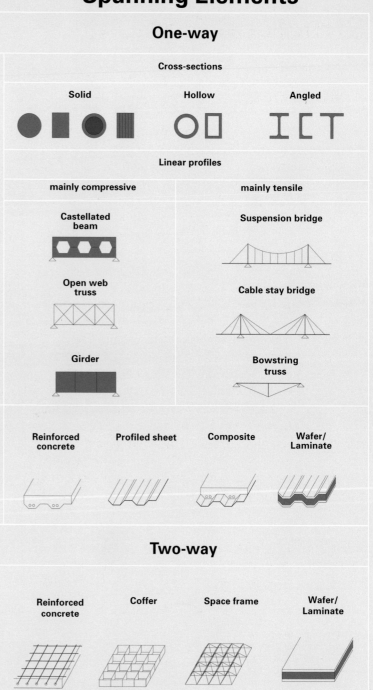

One-way

Cross-sections

Solid

Hollow

Angled

Linear profiles

mainly compressive

mainly tensile

Beam

Castellated beam

Open web truss

Girder

Suspension bridge

Cable stay bridge

Bowstring truss

Profiled plate

Reinforced concrete

Profiled sheet

Composite

Wafer/ Laminate

Two-way

Reinforced concrete

Coffer

Space frame

Wafer/ Laminate

Structural Elements

With all structural forms, the size, number and location of the different structural elements will be choices made by the designer as part of the overall design concept. The designer must be able to describe how the different elements work individually, and how they contribute to the way the building works as a uniform structure. Structural elements are classified as either bearing – those that carry loads to the ground mainly in compression (such as columns and foundations) – or spanning, i.e. those elements that must span or bridge space (such as floors and roofs).

Bearing Elements

A structure made with solid walls has the advantage that the walls also brace the structure – you could think of them as extended columns in a pinwheel. These walls can support floors and roofs, and are known as loadbearing walls. All other (non-loadbearing) walls are either partitions when used internally or weathering/insulating screens when used externally (see Climatic Envelope, pages 80–101). Loadbearing walls create a climatic enclosure but at the same time have the disadvantage that they carry their loads to the ground throughout their length and hence require a supporting beam or continuous strip foundation (see Foundations, page 42).

Floors and roofs may also be carried on a series of columns each of which will have its own, independent foundation. A loadbearing wall may also be reinforced by embedding columns into the wall at various intervals. These are known as engaged columns.

Foundations

Foundations are required for structures that carry loads to the ground, and are usually created by filling holes dug into the ground with heavy material such as concrete or stone. Alternatively with soft ground, rods – known as piles or piers – can be inserted to the point where stability is achieved either by piercing through to the bedrock, or through friction with the soil itself.

Spanning Elements

Roofs and floors can be constructed in two basic ways, either using multiple, linear spanning elements – beams – or two-way spanning elements – slabs. Roofs and floors may be considered as similar structural elements since they both carry live loads directly. Floors must carry the load of people, furnishings, and equipment while roofs must account for people, equipment, snow, and wind.

Left
Bearing and Spanning Elements
taxonomy charts: the charts categorize
primary bearing and spanning elements
according to their function and type.

36 Bearing Elements / Bearing Walls: Stacked Masonry

See also: Mud Hut p134
Bell Rock Lighthouse p146

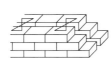

Masonry

Loadbearing walls are traditionally made from piling up bricks or blocks made from dense, massive materials such as clay, stone, or concrete. Such walls are referred to as masonry walls. Each layer of blocks is known as a course, and these are laid in patterns so that joints in between them do not run vertically through the wall. These patterns are known as the bond. Stone walls, particularly where they are used as boundary walls, may be constructed without mortar, relying entirely upon gravity for their stability; these are known as drystone walls. Most modern loadbearing masonry, however, is bonded together using a ⅜ in bed of mortar, which is a mix of soft sand and cement (usually at a ratio of 4 : 1) and water. Adding lime to the mortar allows the mix to remain plastic so that it can adjust to any movement in the wall and prevent cracking.

Clay Bricks

These are made from extruding or pressing clay into a mold and baking the bricks in a kiln. Bricks are often made with a recess (known as a frog) or with 1 in holes running through them so that the mortar binds the bricks in such a way as to restrict any linear movement. A standard brick is 8 x 3 ½ x 2 ½ in. The small face of a brick is known as its header, and the long face as its stretcher.

Adobe Bricks

Adobe bricks are made from mud (for compression) and straw (for binding – tension) and are baked in the sun. The walls are then usually rendered by plastering mud onto the surface to improve cohesion and weathering.

Concrete Masonry Units

Concrete Masonry Units (CMU) are generally made as hollow blocks that are laid up in courses bonded by mortar much like clay brick. For additional strength and resistance to lateral loads, steel reinforcing bars may be run vertically through the cells of the block, and the cells grouted solid. The standard size for CMU is 8 x 8 x 16 in.

Thickness of Masonry Walls

Traditional stone walls were often very thick – not only to carry floors and roofs, but also to protect the occupants of the building from outside threat.

For modern, two- or three-story structures, however, an 8 in thick wall is usually structurally sufficient. The better masonry bearing walls built today employ an interior wythe of CMU that carries the structural loads, a cavity (for insulation, waterproofing, and drainage) , and an outer masonry veneer as exterior finish.

1 Flemish-bonded brickwork.
2 Cavity wall.
3 Reinforced blockwork.
4 **A**, Alternate courses of Flemish-bonded brickwork. **B**, Alternate courses of English-bonded brickwork. **C**, Plan of cavity wall: two skins of brick or blockwork are tied together with a gap of around 4 in – for insulation and to prevent moisture penetrating – in the middle. Metal ties are embedded into the mortar beds at roughly 1 ft 6 in intervals, horizontally and vertically.
5 Flemish-bonded loadbearing brickwork.
6 Traditional loadbearing stonework.
7 Machu Picchu, Andes, Peru. Loadbearing stonework cut with great accuracy by hand and laid dry, i.e. using no mortar.
8 Loadbearing cavity wall. The outer skin is brickwork, the inner skin artificial blocks. One block is the size of six bricks.

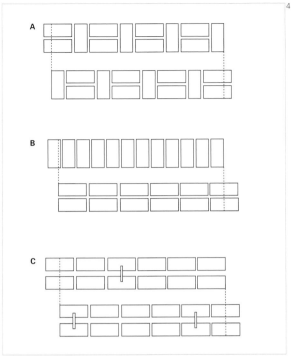

A

B

C

38 Bearing Elements / Bearing Walls: Composite Construction

1

Cast Concrete

As well as being made by stacking up modular blocks, bearing walls may also be constructed through casting or molding "state-change" materials – those that can be poured into a mold and that will then solidify and take the shape of that mold through one of a number of chemical processes. The most common method of casting a loadbearing wall is by using reinforced concrete. Wet concrete is poured into a mold that contains steel reinforcing bars. These types of bearing wall may be constructed off-site in large, modular panels – a process known as precasting.

Additional Composites

Loadbearing walls can also be constructed using materials within a composite framework. Sandwiching panels around an inner core or framework results in a form of construction known as monocoque or stressed skin. The two outside panels act in tension to stiffen the inner core.

1 Composite bearing walls: reinforced concrete and two stressed-skin varieties – framed and laminated.

2 In this instance, concrete is pumped into a formwork consisting of prefabricated hollow, polystyrene blocks which are left permanently in place to insulate the building. The outside surface is then weatherproofed.

3 Concrete has been poured into timber shuttering. The shuttering can be removed – struck – once the concrete has set hard, usually after 24 to 48 hours. This is known as the curing process, and concrete will reach 75 per cent of its full strength in around seven days, and will only reach maximum strength after a period of weeks.

4 Reinforced-concrete wall, cast on site – in situ. As well as bearing the floor loads, the wall is acting to resist lateral loads – to brace the structure. While concrete is very strong in compression, it needs reinforcing with steel bars to resist tensile forces.

5 Loadbearing concrete wall, showing reinforcing bars.

6 Cast-concrete loadbearing wall. The holes are where the two sides of the shuttering were tied together to prevent them being forced outward by the weight of the wet concrete. The surface of the concrete takes the imprint of the shuttering material – in this case timber planks.

7 Reinforced-concrete walls, cast in situ to support a stairwell.

8 Prefabricated structural insulated panel (SIP) made by bonding oriented strand board (OSB) – to a foam core, a process known as lamination.

2

3

4

5

6

7

8

See also: The Crystal Palace p148
HSBC Headquarters p168

1

Walls and roofs can be supported on columns (also known as pillars or piers) as well as by loadbearing walls. The intervals between columns in a grid is directly related to their cross-sectional size – the smaller the section, the closer together they will be. Assuming a 26 ft 3 in by 26 ft 3 in grid, columns can be roughly sized (cross-sectional area) on the basis of the number of floors supported:

1 story	1 ft x 1 ft
6 storys	2 ft 3 in x 2 ft 3 in
11 storys	3 ft 3 in x 3 ft 3 in
16 storys	3 ft 11 in x 3 ft 11 in

Buckling

The length of a column will determine the way in which it will react under load. The taller the column, the more likely it is to buckle (bend or "kneel"), and its stiffness rather than its compressive strength becomes important. It is for this reason that a tubular section column is more efficient than a solid section one.

Loads

The loads in a column gradually increase toward the base. To compensate for this, columns may taper or get narrower towards the top.

Classical Orders

Ancient building types are classified by the type of column used in their construction. Known as the Classical Orders, columns are divided into Doric, Ionian, and Corinthian – all of Greek origin – Tuscan and Composite, which are of Roman origin. A classical column is described using three main elements: its shaft, its base and its capital, and the different orders may be distinguished by the detailing and proportioning of these. The height of a column is measured as a ratio between the diameter of the shaft at its base compared to its height.

For example, a Doric column can be described as being seven diameters high, an Ionic column as eight diameters high and a Corinthian column as nine diameters high.

1 Cross-sections of columns. From left to right: four examples of solid columns, two of hollow columns and one H-section column.

2 Relative column sizes in a 26 ft 3 in square, 12-story tower.

3 The Parthenon, Athens, Greece. The stone columns are made in sections. They are Doric in style and are proportioned in the ratio of 6:1 – roughly the length of a man's foot in relation to his height. They are 33 ft high, have a 5 ft 7 in diameter and are spaced at 13 ft centers.

4 Columns at Le Corbusier's Villa Savoye built, in 1928, are made from 1 ft diameter reinforced concrete.

5 Stone columns at the Sagrada Família church, Barcelona, Spain, by Antonio Gaudí.

6 The intervals between columns in a grid are directly related to their cross-sectional size. The Aluminum Knowledge and Technology Center ("The Aluminum Forest"), Utrecht, the Netherlands, by the architect Micha De Haas is supported by 368 aluminum tubes. The tubes are 19 ft 6 in high and have a diameter of between 3 ½ and 8 in.

7 Preformed tubes used as shuttering for circular section columns.

8 Branching columns at the Royal Ascot Racecourse, Windsor, UK, by the architects HOK Sport.

9 Inclined concrete columns at the Palestra building by SMC Alsop Architects. Angling and raking columns can help to brace a structure.

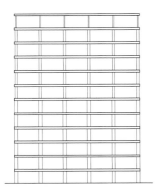

2

3

4

5

6

7

8

9

42 Bearing Elements / Foundations: Strip, Pad, Raft, Pile, Retaining Walls

See also: Bell Rock Lighthouse p146
Self-build House p164
Davies Alpine House p180

1

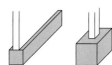

Foundations occur where structures carry loads to the ground. They are usually formed by filling holes dug into the ground with dense, massive material such as concrete or stone, or alternatively by driving rods – known as piles – into soft ground to the point where they become stable either through friction or through reaching down to bedrock.

Types

The stability and topology of the terrain will determine the type of foundation that is needed. Not only will the foundation prevent the structure sinking into the ground, it will also provide a stabilizing anchorage – a fixing. Shallow foundations (sometimes called spread footings) include strip footings, pads (isolated footings) and rafts. Deep foundations employ piles and retaining walls.

Shallow Foundations

For a bearing wall, a continuous or strip foundation is needed to support it throughout its length. An individual "point" load such as a column will require a pad or pile foundation. Raft foundations are used on unstable ground, and are so named because they act like a single, rigid element that "floats" on the terrain. Often it is necessary to use various foundation types in combination, for example a series of piles may be connected with a ground beam which acts like a strip foundation to carry a bearing wall.

Deep Foundations

Deep foundations are used where there is unstable terrain and are made by boring deep holes into the ground and inserting piles or piers. Piles can be made from a variety of materials such as timber, concrete and steel. Concrete piles may be precast or cast in situ and steel piles are usually either hollow tubes (pipes) or H-sections. Common, cast in situ concrete piles may be made either by pouring concrete into a drilled hole (which may contain steel reinforcement) or by driving a temporary or permanent steel casing into the ground and filling it with concrete.

Retaining Walls

When ground preparation also requires the excavation of material it may be necessary to form either temporary or permanent walls to hold back the surrounding terrain. This can be done using cast reinforced concrete (much in the way it is used for a swimming pool), interlocking sheet piles (profiled steel sheets that can be driven into the ground) or diaphragm walling which consists of a series of connected piles (known as secant piles).

1 Foundations/retaining walls. From left to right: strip, pad, raft, pile, retaining walls
2 To restrain a tensile load, either a friction anchor (similar to a tent peg) or mass is used.
3 Concrete cast in trenches to form strip foundations for a loadbearing wall.
4 Pad foundations carry loads at points.
5 A concrete raft foundation reinforced with a steel mesh. This will spread loads evenly over a surface.
6 Bored pile with steel casing. Concrete will be poured around the reinforcing bars.
7 Timber shuttering for a cast in situ, reinforced-concrete retaining wall.
8 Tensile connections to a concrete foundation.
9 Secant pile retaining wall. This is made by boring holes adjacent to one another and filling them with concrete. Excavation can then take place to reveal a series of "engaged" concrete piles (also known as contiguous piles). A vertical reinforcing mesh has been applied to the face of the secant piles; this will be sprayed with concrete to form a composite retaining wall with a continuous surface.
10 Sheet piles. Interlocking steel sheets are driven into the ground by a machine.

⁴⁴ One-Way Spanning Elements / Beam: Cross-Sections

1

Beams are used as multiple, unidirectional spanning elements to carry floor and roof plates. Once a surface plate has been fixed to the beams, the whole floor or roof becomes a rigid (horizontally braced), structural element. Beams are described using a combination of their cross-section and linear profile.

Cross-Sections

The primary purpose of most beams is to resist bending forces. Bending induces compressive forces on the top of the beam and tensile forces on the bottom in a simple span from one element to another. The more beam material that is concentrated away from the member's centerline, the more resistance to those forces that section will have. Therefore, beams should have a greater depth than width, and rectangular tubes or "I" shaped sections make more efficient use of material than solid rectangular sections.

Most beams maintain a consistent cross section across their span even though stresses vary in type and location in the beam along that length. Material may be used more efficiently by adjusting the depth or section of the beam in response to the stress that it receives at each point in its span. However, this is generally more expensive and is only done in specialized circumstances.

A beam-like spanning member used repetitively is referred to as a joist. Joists may be of many different materials and are generally lighter in weight than beams.

1 Cross-sections of beams. From left to right: four examples of solid beams, two of hollow beams and three of angled beams.

2 Extruded steel wide flange beam. The web resists bending and the flanges resist shear and torsion. Each flange tends to be between half and one-third the size of the web. Steel beams are specified according to weight per foot, area of section, and web and flange thickness. **As a general rule, rolled steel sections have a span/total depth ratio of around 20 : 1.**

3 The Parthenon, Athens, Greece. The span between the columns is only 7 ft 6 in, nevertheless this stone beam has cracked and been reinforced with metal pins to resist the tension low down in the beam. A beam that spans to form an opening in a wall is known as a lintel.

4 Wood roof joists. These are 1 ½ by 7 ½ in spaced at 2 ft on center. They span 14 ft 9 in. **For average residential floors with joists spaced at 1 ft 4 in centers, the required depth in inches of a 2x joist is approximately half the span in feet plus two inches.**

5, 6 Glued laminated timber beams (Glulam). Pieces of timber are glued together to create strong, versatile beams that can be shaped and curved. As a general rule, glulam beams have a span : total depth ratio of around 18 : 1.

7 Timber I-joists. These comprise a timber flange, typically solid timber and a panel product web, usually oriented strand board (OSB).

8 Glass: these structural glass beams comprise three pieces of ½ in. toughened glass laminated together.

9 Concrete T-section beams. Used in conjunction with smaller concrete slabs. **As a general rule, rectangular section, reinforced-concrete beams have a span : total depth ratio of around 23 : 1.**

10 Concrete box girder. Made up from precast concrete box sections; the upper flange acts as a deck.

11 Pneumatically-inflated air beams stitched together to form a canopy.

See also: Clifton Suspension Bridge p150

1

2

      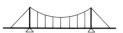

Linear Profiles

While a simple beam will maintain the same cross-section throughout its length, greater spans require the removal of material that is structurally redundant – material in the beam that is not working to carry loads – in order to reduce the dead load of the beam itself. This arrangement will tend to create deeper beams but (being relatively lighter) they are able to achieve long spans. Beams can be modified in this way by cutting holes in the web (as with a perforated or castellated beam) or by replacing a solid web with a series of vertical and diagonal struts (as with an open web truss; usually used for spans of over 50 ft.

Pure Tension

While compression structures commonly employ elements that are working in both tension and compression (see Forces: Stress, Strain, page 16), tensile structures also employ elements that are purely in tension. Long-span structures typically employ substantial tensile elements in order to reduce the dead loads to the minimum and are counterbalanced and anchored at each end.

1 Linear profiles. Mainly compressive: solid beam, castellated beam, open web truss, and girder.
2 Linear profiles. Mainly tensile: bowstring truss, cable stay bridge, suspension bridge.
3 Process of forming castellated beams. Redundant material is removed from the web and, in the process, the depth of the beam is increased. These types of beam are capable of spanning up to 66 ft.
4 Open web trusses have diagonal cross-bracing to prevent them failing under eccentric load conditions. (Holed and castellated beams maintain triangulation in the body of the web.)
5 Perforated steel I-beams.
6 Composite open web joist: wood with steel connectors.
7 Open web trusses made from cold-rolled steel.
8 Open web truss as a pitched roof, known as a trussed rafter.
9 Plate girder. An I-section beam with a solid web composed of individual steel plates and vertical angles that are welded, bolted or riveted together.
10 Vierendeel truss. A truss that has no diagonal members but braces the frame using triangular connections at every junction between an upright and the top and bottom rails.
11 Cable stay bridge with cast-concrete masts. The platform is hung via cables that radiate from either side of the columns; the load is thus balanced through a double cantilever.
12 Clifton suspension bridge, Bristol, UK. Suspension bridges span long distances by hanging cables (in this case wrought-iron chains) between tall masts. The road decks are then suspended from vertical cables set at regular intervals.

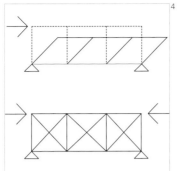

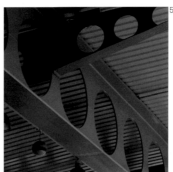

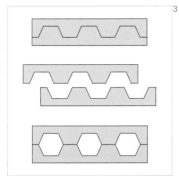

48 One-Way Spanning Elements / Profiled Plate: Reinforced-Concrete Slab, Profiled-Sheet, Composite, Wafer/Laminate

1

In addition to beams, modern materials and methods of construction enable floors and roofs to be formed as single, monolithic plates.

Floor and roof plates that are formed as a single slab or "decking" can be designed to span either in one direction (one-way systems) or multi-directionally (two-way systems). There are two main ways of stiffening a one-way plate. One is to design in dropped beams to stiffen the plate and the other is to fold the plate into a corrugated profile.

As with a beam, a profiled slab or deck is a linear spanning element that maintains its cross-section throughout its length, but unlike a beam, it forms a continuous surface. Typically made from poured or cast materials such as concrete or plastic, profiled decks can also be formed using folded sheet metal – or by laminating profiled sheets around a core.

Span to Depth Ratio
As a general rule, the ratio of span to total depth for a continuous reinforced-concrete slab is around 32 : 1, and for a coffered slab, around 26 : 1. A steel space frame will have a span to depth ratio of around 15 : 1.

1 Folded plates. From left to right: reinforced concrete, profiled-sheet, composite, wafer/laminate.
2 Profiled metal sheets (metal decking) are used as permanent formwork and act partially to reinforce the in situ poured concrete.
3 Profiled metal decking is made from galvanized steel formed into a trapezoidal section.
4 Folded sheet. The principle employed for profiled metal sheets is used at a larger scale whereby spanning is achieved by casting a continuous concrete slab with a profiled or folded section.
5 Corrugated or profiled metal sheets are commonly used for small-span roofing.
6 Cast-concrete slab with embedded down-stand beams. These down-stands contain reinforcing bars so that the slab can resist tensile forces.

50 Two-Way Spanning Elements: Reinforced-Concrete Slab, Coffer, Space Frame, Wafer/Laminate

1

   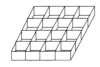

A two-way slab or deck is a surface that can act like a single, rigid element and be supported at a variety of points. These elements may be cast as a single component, such as a solid or coffered, reinforced-concrete slab. They may be formed by laminating sheets around a core (as in a wafer), or assembled using a framework of smaller elements as with certain types of coffer or as a space frame (also known as a space grid).

Space Frame

In the same way that a truss may be considered as a series of linearly connected triangles, a space frame is a grid of connected pyramids or tetrahedra. The relative lightness of this type of structure means it can achieve greater spans than other types of two-way spanning elements.

Span to Depth

As a general rule, the ratio of span to total depth for a continuous reinforced-concrete slab is around 32 : 1, and for a coffered slab around 26 : 1. A steel space frame will have a span to depth ratio of around 15 : 1.

1 Multidirectional spanning elements. From left to right: reinforced concrete, coffer, space frame, wafer/laminate.

2 Composite slab. A cast-concrete slab has an internal mesh of steel reinforcement to resist tensile forces.

3 Concrete reinforcement consists of steel bars joined loosely together to form a three-dimensional cage.

4 Cast-concrete slab showing the surface imprint of timber formwork.

5 Space frame. A lightweight, three-dimensional grid consisting of a series of linked pyramid shapes.

6 Coffered floor slabs.

7 By placing hollow plastic spheres and ellipsoid shapes within the steel reinforcement mesh, overall deadweight is reduced making larger, more efficient spans achievable.

8 Detail of plastic spheres in the steel reinforcement "cage". These slabs are a 3D version of a castellated beam. Where the concrete is not carrying loads it is replaced with air.

9 Laminate. A slab can be made up of layers that are bonded together, such as this example of aluminum plates bonded to a honeycomb aluminum interior. Glass-reinforced plastic (GRP) may also be used to form structural elements by laminating foam cores with glassfiber-reinforced resin (yacht hulls are often made in this way).

10 Coffer. A 6 ft deep grid of welded steel girders forming the roof of the New National Gallery, Berlin, Germany by Mies van der Rohe; the 212 by 212 ft roof is supported by eight columns.

52 Connections and Joints: Dry Joint, Stone, Nail, Screw, Bolt

To understand how building components and building elements are assembled, it is necessary to explore the various methods of fixing and joining different materials.

Dry Joint: a pressure joint
The term joinery refers to the age-old tradition of connecting pieces of timber together to form a whole. Dry timber joints rely solely upon the precise geometry of the cuts made in the timber and the plasticity of the timber for a really close fit. Many connection types are possible. With a scarf joint, for example, the connection is designed so that two pieces of timber can be joined longitudinally, and so that the resulting element is able to resist tension, compression and torsion. To complete the joint, timber dowels (circular-section rods) are inserted into pre-drilled holes; in order to lock these dowels in position a hardwood wedge is cut to the same width as the dowel and then hammered into a saw cut.

Stone
Reference has already been made elsewhere in this book to drystone walls. Stone is cut and laid with an accuracy and in such a way as to rely solely on gravity for the integrity of the structure.

Nail: a friction joint
Nails are known to have been used to connect timber as long ago as the ancient Roman period. A nail is essentially a smooth shaft with a sharp point at one end and is typically made from steel. Nails are driven into wood by a hammer or a nail gun driven by compressed air. A nail holds materials together by friction in the vertical direction and shear strength in lateral directions. Nails are made in a variety of forms according to purpose; they can have a flat head to prevent them penetrating the surface of the wood or they may have a small oval head that will sink into the wood. Nails are normally cylindrical but "cut nails" (machine-cut from flat sheets of steel) have a rectangular cross-section. Common nails are known as wire nails.

Screw: a removable friction joint
A screw consists of a metal rod that has a head on one end and a continuous helical thread on the other that tapers to a point. The head may come in a variety of shapes according to function. However, all will contain a recessed groove or cross so that they can be turned using a screwdriver. A screw fixing relies on friction with the material into which it is bored and is most often used with wood.

Bolt: a removable connector
A bolt consists of a metal rod that has a head on one end and a continuous helical thread on the other. It is used in conjunction with a nut that contains the negative imprint of the thread. The bolt is passed through a pre-drilled hole in two or more objects and the nut is tightened over the protruding end. Both the bolt head and the nut are usually hexagonal in shape so that they can be turned using a wrench or spanner. Bolts are normally tightened to a precalculated load known as the torque load.

1 Scarf joint assembly.
2 Common wire nail.
3 Woodscrew.
4 Basic bolt with a hexagonal head.
5 Dovetail joint.

Firstly, the timber sections to be joined are cut to identical profiles; these are then connected on opposing faces using "butterfly" timber wedges; the whole assembly is then locked together using a timber dowel – the dowel is fixed by hammering hardwood wedges into grooves at either end.

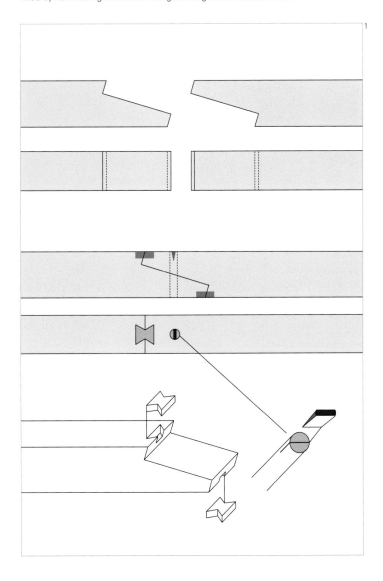

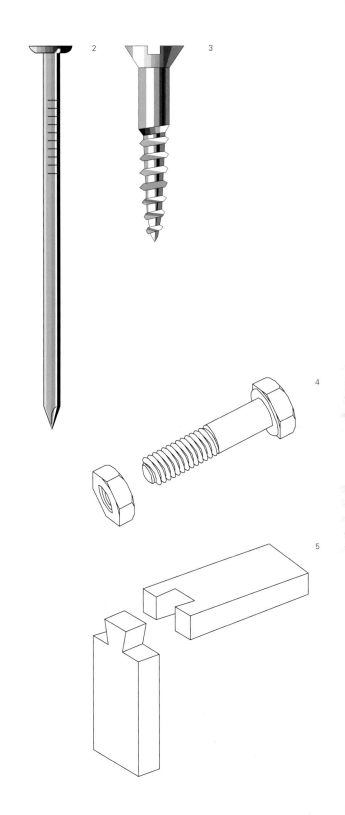

54 Connections and Joints: Rivet, Turnbuckle, Adhesives (Glue), Welding, Brazing

Rivet: a one-sided joint

A rivet is a mechanical fastener. It starts off as a smooth cylindrical shaft with a head on one end. The rivet is driven into a pre-drilled hole so that the shaft becomes deformed and expands to about one and a half times the original diameter. Original "solid rivets" required two assemblers: one person with a hammer on one side and a second person with a bar on the other side. Unlike solid rivets, however, modern "blind rivets" can be inserted and locked into position from only one side of a structure. They are formed as a tube with a mandrel through the center, and a specially designed tool is used to draw the mandrel into the rivet, expanding the rivet and snapping the mandrel.

Turnbuckle: a tensioning mechanism

A turnbuckle is a device for adjusting the tension of ropes, cables and tie rods. It normally consists of two threaded eyelets, one with a left-hand thread and the other with a right-hand thread, both screwed into each end of an extruded nut. The tension can be adjusted by rotating the nut, which causes both eyelets to be screwed in (or out).

Adhesives (Glue)

An adhesive is a liquid compound that bonds two items together and may come from either natural or synthetic sources. There are adhesives that can be used on almost any material or combination of materials. Adhesion may occur either by mechanical means, where the liquid adhesive penetrates into the material and then hardens, or by one of several chemical mechanisms. The first adhesives came from plants (resins) and animals (fat).

Welding

Welding is the only way of joining two or more pieces of metal to make them act as a single piece. With welding, metals are joined by applying heat, sometimes with pressure and sometimes with an intermediate, molten filler metal. By applying intense heat, metal at the joint between two parts is melted and caused to intermix. Once cool, welds can be ground flat using an abrasive wheel. The most common methods for fusing metals in this way are arc welding and metal inert gas (MIG) welding. With arc welding, the intense heat is produced by an electronic arc. The arc is formed between the metal and an electrode (rod or wire) that is manually or mechanically guided along the joint. Often, the rod or wire not only conducts the current but also melts and supplies filler metal to the joints. With MIG welding, an aluminum alloy wire is used as a combined electrode and filler material. This semi-automatic process feeds a continuous spool of filler wire to the weld bead by magnetic force – a process known as spray transfer, which enables welding from any position.

Brazing

While in welding the metal of the joining surfaces is fused by melting, brazing requires the joining surfaces to be heated, but only the filler material to be melted. The filler metal has a lower melting point than the metal(s) to be joined. It joins by bonding rather than fusing. It generally employs lower temperature than would be required to join the same materials by welding. More accuracy is required with brazed joints than with welded connections. However, if the geometry of the faces to be mated is well matched a brazed joint will be stronger than the parent metal.

1 MIG-welding aluminum sheet metal.
2 Arc-welding structural steelwork.
3 Bolted steel connection.
4 A "blind rivet".
5 A "reverse threaded" turnbuckle.

1

2

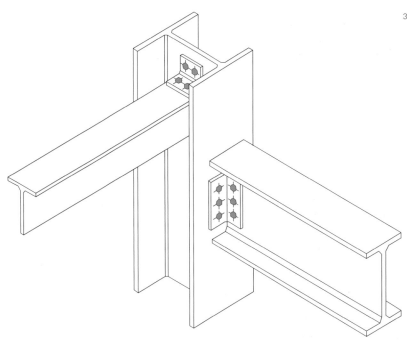

3

4

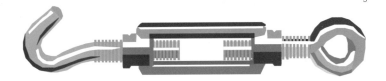

5

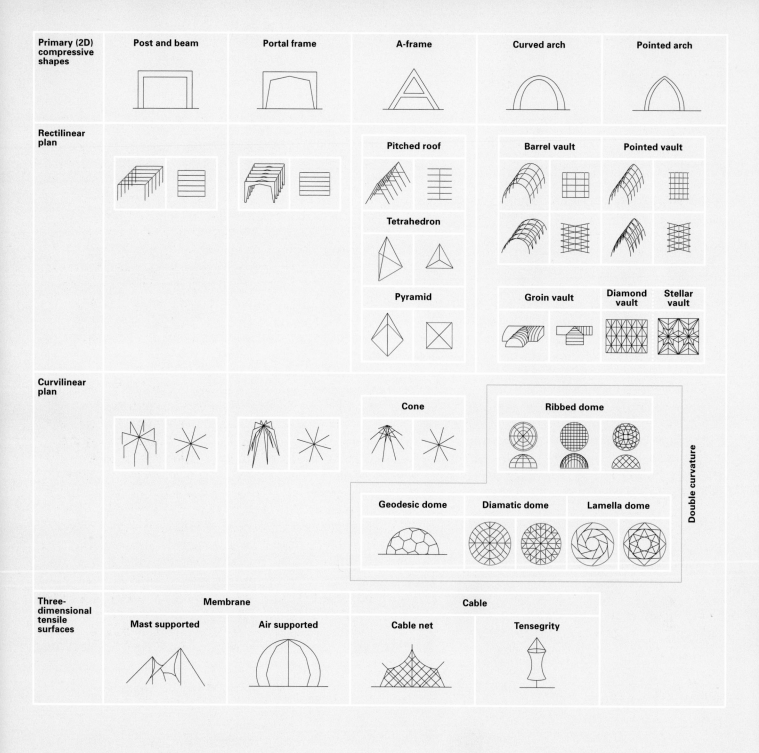

57 Structural Logic

The term structural logic is used to describe how structural elements are combined to form different building types. Structures are classified here as mainly compressive or mainly tensile, and are geometrically described as either a **two-dimensional shape** that is used repeatedly (iterated) or as a **three-dimensional (double-curved) surface**.

It is considered that, to some degree, any of the construction elements described in the previous section can be applied to any type of structural form. Because of this, the illustrations in this section attempt to cover not only different types of structural logic but also a variety of construction solutions and the particular structural elements that they employ.

Two-Dimensional Shapes

In two dimensions, compressive structures can enclose an area in four basic ways: as a beam and post, an A-frame, an arch or a portal frame. These two-dimensional elements or shapes may then be iterated – used repeatedly – to create volumes through extrusion and/or rotation. A series of building types derive their forms from the iteration of two-dimensional shapes, and these are outlined in this section. Derivations of the arch have been a particular source of creative structural logic, and these are covered separately.

Surfaces

Certain structural forms cannot be described by applying a simple geometric procedure to a curve or arch. Geometrically, spheres, cylinders and cones may be described as surfaces that enclose (or partially enclose) space. In a similar way, cubes, parallelpipeds and other polyhedra may be described as surfaces. A surface may be described as a two-dimensional geometric figure (a collection of points) in three-dimensional space. Mathematically, surfaces are normally defined by one or more equations, each of which gives information about a relationship that exists between coordinates of points on the surface, using some suitable (e.g. Cartesian) coordinate system. A series of building types derive their forms from surface geometry, and these are outlined in this section.

Tensile Structures

Geometrically, a tensile structure is normally described as a surface. Such structures are constructed by stretching fabric (membranes) or a network of cables (cable nets) using masts or other shapes to carry the compressive loads and ground anchors to resist the upward pull. For these surfaces to be structurally efficient – where the whole surface is in equal tension and there is no structurally redundant material – there tend to be certain geometric shapes. Mast-supported membrane and cable net structures develop saddle type or anticlastic shapes – they consist of two curvatures (hyperbolic paraboloids) that traverse in opposite directions. Air-supported membrane structures tend to develop synclastic shapes – they curve towards the same side in any direction, as with a sphere.

Left
Structural logic taxonomy chart.
Red lines indicate bracing elements.

58 Compressive Structures / Two-Dimensional Shapes: Post and Beam, A-Frame, Portal Frame, Arch

1

Post and Beam

The simplest structure is where two vertical columns support a beam, which spans between them.

A-Frame

This structure spans through triangulation, with each side leaning against the other. If its legs are not fixed at their lower end, the structure needs horizontal bracing that is mainly in tension to prevent the legs from splaying out under load.

Portal Frame

A portal frame is a combination of a post and beam and an A-frame. It works in a similar way to a Vierendeel truss, being composed of a rigid frame that is made up of elements with a deep cross-section to help brace the structure.

Arch

An arch may be curved or pointed. The curved arch is a continuous compressive element that distributes loads through its curvilinear form. The loading in an arch increases towards its base, which must be anchored or tied to prevent the arch from splaying outwards.

Catenary, Parabolic and Elliptical Curves

Each of these curves is considered to be structurally efficient when employed as an arch, the elliptical arch spanning longer distances than standard arches due to its flatter curve. A catenary curve may be derived from inverting the profile of a hanging chain. Parabolic curves also occur naturally, as in the profile of a water jet, and both parabolic and catenary curves can be expressed as a quadratic equation derived from plotting them on a graph. In a suspension bridge, the cables that are stretched between the masts form a catenary curve, however once the cables become loaded (by hanging a deck from vertical cables placed at regular intervals) the curve becomes parabolic. When a catenary curve is inverted, it forms a naturally stable arch. Arches formed in this way are structurally efficient since the thrust into the ground will always follow the line of the arch.

The parabola, in its simplest form, is:
$$y = x^2$$

The catenary is defined by the hyperbolic cosine:
$$y = \cosh(x) = (e^x + e^{-x})/2$$

Pin-jointed Arches

The pin joint is a structural connection that allows for adjustment through rotational movement at the point of connection, similar to a hinge. Arches employ pinned joints at the base and/or apex, partly for ease of construction and partly to allow for differential movement. If the arch were to be fixed rigidly, stresses would build up whenever it became subject to non-uniform external loading, such as ground movement.

1 Primary (2D) compressive shapes. From left to right: post and beam, A-frame, portal frame, curved arch, pointed arch.
2 Catenary curves.
3 A single-pinned arch will allow for limited movement at ground level. A 2-pinned arch allows for one side or the other to act as a pivot point. Adding three hinges allows two halves to lean against each other, offering multiple degrees of freedom of movement.
4, 5, 6 Post and beam.
7 Portal framing system by Lamisell Ltd, UK.
8 A-frame.
9, 10 Although these structures designed by Nox Architecture are arching, the spanning principle is that of a portal frame.
11 A parabolic curve may be drawn by cutting any oblique section through a cone. The intersection of a cone with a plane results in an ellipse.

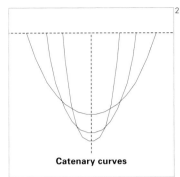

2

Catenary curves

3

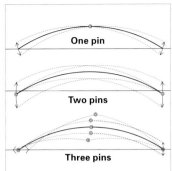

One pin

Two pins

Three pins

5

6

7

8

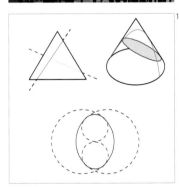

9

10

11

60 Compressive Structures / Two-Dimensional Shapes: Vault, Dome

See also: Pantheon p142
Crystal Palace p148

1

Barrel Vault

A barrel (or wagon) vault is formed by extruding an arched shape. It maintains a constant cross-section throughout its length.

Groin Vault

A groin vault is the result of two barrel vaults intersecting at right angles. The line of intersection of vaulting surfaces is an arched diagonal known as the groin.

Gothic Vault

While a traditional Romanesque arch was a curve, Gothic arches borrowed from the Muslim tradition and were pointed. Gothic roofs also combined diagonal ribs with the groin vault (although it is still not known whether they helped support the vault). Variations include the diamond vault, the stellar vault and the fan vault.

Ribbed Dome

A ribbed dome is formed by rotating a set of identical arches about a central axis.

1 From left to right: two examples of barrel vaults, two examples of pointed vaults, three diagrams of ribbed domes.

2 Barrel vault made from curved, iron trusses. The Natural History Museum, London, UK, designed by Alfred Waterhouse and completed in the late nineteenth century.

3 Brick groin vault at the Natural History Museum.

4 Ribbed dome. The horizontal members act as a continuous ring that holds the ribs in tension and prevents them spreading outwards under the load. The dome is made from curved, laminated timber beams, fabricated by bending a strip of timber and then gluing successive layers or strips on top of each other; each layer maintains the curved shape of the one below.

5 Downland Gridshell: patented joint used in construction.

6 Downland Gridshell, Weald and Downland Museum, Sussex, UK, Edward Cullinan & Partners with Buro Happold and the Green Oak Carpentry Company. These lightweight, timber lattices need to be tensioned around their edge using a continuous ("ring") beam.

7 Fan vault at Wells Cathedral, Somerset, UK, built largely in the late twelfth and early thirteenth centuries.

8 Le Galerie des Machines was built for the 1889 Paris exhibition. Designed and engineered by Duter, Contamin, Pierron and Charlton, the hall was formed by a series of three-pinned, trussed, steel portal frames.

9 Berlin Hauptbahnhof, Berlin, Germany, by Gerkan Marg and Partners.

10 Orangery, Prague Castle, Czech Republic, designed by Eva Jiricna and engineered by Mathew Wells of Techniker.

62 Compressive Structures / Three-Dimensional Surfaces: Diamatic Dome, Lamella Dome

See also: Palazetto Dello Sport p156
The Houston Astrodome p158
The USA Pavilion p162

1

Diamatic Dome

Spans may be increased by using this system which combines the arch with the "A" frame. The primary structure is made from any number of intersecting arches with lateral rings, as with a conventional ribbed dome. The arches are then reinforced with triangular bracing.

Lamella Dome

This dome employs a helical arrangement of ribs, which when used in opposing directions act to brace as well as support the structure. The dome is generated with concentric rings, where each subsequent ring is rotated by a half module. This reduces the length of the ring tubes as the geometry proceeds towards the apex. The separation between rings in Lamella domes can be varied so they are equilateral triangles forming each ring.

The St. Louis Arena was one of the first large scale lamella domes constructed in the US, which opened in 1929 (demolished 1999) and was designed by Dr Gustel R. Kiewitt. Kiewitt created this huge roof, 145m long and 84m wide, using small Douglas fir timbers to create a triangulated "fish-scale" structural shell. Kiewitt was also involved in the design of the Houston Astrodome (1962–64), with a span of 196m. The same structural principles were subsequently applied in the construction of the recently refurbished New Orleans Superdome (1973), 207m in diameter.

The lamella principle of a crisscrossing pattern of short structural members

(lamellae) interlocking in a diamond pattern can also be used to form a vaulted roof structure. Hugo Junkers (1859-1935), more famous for his aeronautical innovations, patented a steel lamella roof construction system in the early 1920's, based upon the earlier timber lamella roofs of Fritz Zollinger. Two examples of Junkers lamella system utilizing lightweight pressed metal components were built as aircraft hangars at Rodmarton, Gloucestershire (1938-1939). These innovative parabolic arch structures have subsequently been listed, siting similarities with Luigi Nervi's lamella-type structures built for the Italian air force.

1 From left to right: diamatic dome, lamella dome.

2 The Platonic solids. In geometry, a Platonic solid is a convex, regular polyhedron of which there are no more than five. The name of each solid derives from the number of faces, i.e.
a Tetrahedron (four triangles)
b Cube (six squares)
c Octahedron (eight triangles)
d Dodecahedron (12 pentagons)
e Icosahedron (20 triangles)
The solids are unique in that the sides, edges and angles are all congruent.

3 The Houston Astrodome, Texas, USA (see also p.158). A diamatic dome.

4 The Palazzetto Dello Sport, Rome, Italy, Pier Luigi Nervi (see also p.156). A lamella dome.

5 30 St Mary Axe, London, UK, Foster & Partners (see also p.176). A structure whose surface geometry is derived from a lattice of double helixes, similar to a lamella dome, also known as a tubular or exo-skeletal structure.

6 The roof over the Great Court, British Museum, London, UK, Foster & Partners. On plan the roof consists of a series of radiating lines that are either perpendicular or tangential (like bicycle spokes) to the reading room dome in the center. Since the dome was not in the center of the quadrangle, computer software was designed in order to calculate the optimum adjustments for each facet: there are over 1826 unique (six-way), fully welded connection nodes.

7 Roman Lamella dome – cross-section.

8 Development of a diamatic dome.

9 Structural elements of a lamella dome.

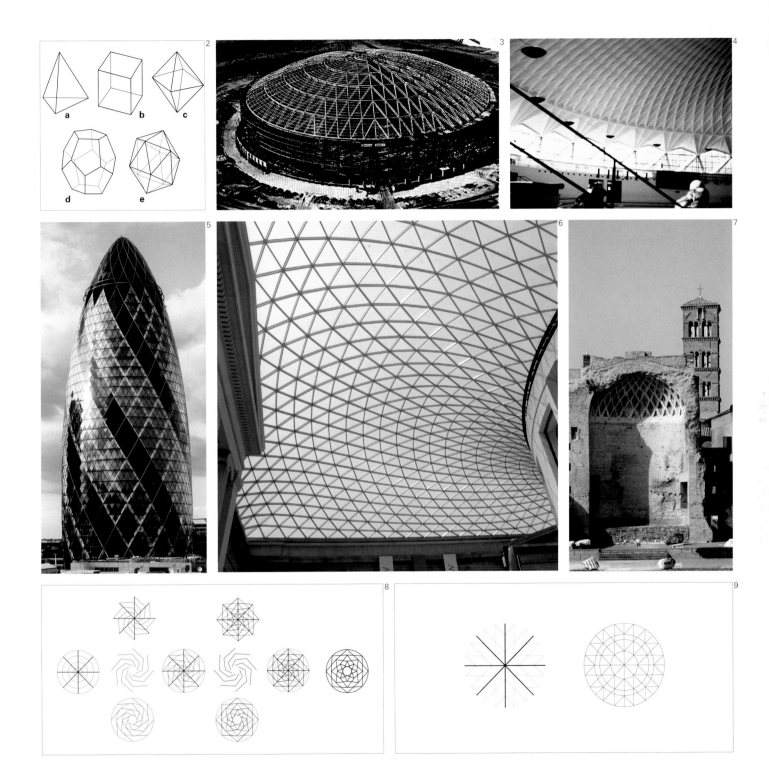

64 **Compressive Structures / Three-dimensional Surfaces: Geodesic Dome**

See also: Palazetto Dello Sport p156
The Houston Astrodome p158
The USA Pavilion p162

Geodesic Dome

The geodesic dome patented by Richard Buckminster Fuller in 1954 is known to be the most structurally efficient of the domes derived from the icosahedron (a twenty sided polyhedron – one of the Platonic or Regular Solids). In the patent application, Fuller described the structure as a spherical mast, which evenly distributes tension and compression throughout the structure. The geodesic dome combines the structural advantages of the sphere (which encloses the most space within the least surface, and is strongest against internal pressure) with those of the tetrahedron (which encloses the least space with the most surface and has the greatest stiffness against external pressure).

A geodesic structure distributes loads evenly across its surface and, as with a space frame, is efficient to construct as it is composed entirely of small elements. The geodesic dome is the product of a geometry based on the shortest line between two points on a mathematically defined surface; it takes its name from the science of geodesy – measuring the size and shape of the earth. A geodesic dome consists of a grid of polygons that is the result of the geodesic lines (or great circles) intersecting. A great circle is (to quote Fuller) "…a line formed on a sphere's surface by a plane going through the sphere's center." And a three-way "great circle" grid can symmetrically subdivide the faces of the icosahedron mapped onto a sphere.

The number of times that you subdivide one of the triangular icosahedra faces is described as the frequency; the higher the frequency, the more triangles there are, the stronger the dome will be. The scalability of the geodesic dome is interesting, with Fuller writing in his seminal publication *Critical Path* "…every time a geodesic dome's diameter is doubled, it has eight times as many contained molecules of atmosphere but only four times as much enclosing shell…" This realization led to Fuller's proposal in 1950 to enclose the whole of midtown Manhattan in a 2 mile diameter geodesic dome whose physical enclosure would have weighed significantly less than the volume of air contained within and whose structure would be largely rendered invisible because of physical proximity and our relative visual acuity.

Fuller and his consultancy companies, Synergetics and Geodesics Inc., produced many structural types of geodesic enclosure working in collaboration with other architects and engineers. Fuller also licensed his technology, which included patented geometric configuration and various connection details. Domes were fabricated from a wide range of materials, which included cardboard, plywood sheets, sheet steel, and Fiber Reinforced Plastics.

1 The Eden Centre, Cornwall, UK. Grimshaw & Partners. A set of interlinking domes (modelled on the highly efficient structural geometry of soap bubbles) made up of hexagonal components.
2 USA Pavilion, 1967 Montreal Expo, Canada, Buckminster Fuller (*see also* page 162).
3 Citizens State Bank, Oklahoma City, Buckminster Fuller with Kaiser Aluminum. Folded shell dome based on the icosahedron.

66 Compressive Structures / Shell and Monocoque

See also: Kresge Auditorium p154
Lords Media Centre p174

Compressive structures may also be designed by utilizing the properties of particular materials. Shell and monocoque structures are two examples.

Cast Reinforced-Concrete Shell Structures

These are traditionally made by pouring concrete into a mold that contains a network of reinforcing bars. You could think of a shell structure as a warped, reinforced-concrete slab. It is usually cast over a timber mold known as the formwork. A type of concrete known as ferrocement can be sprayed onto a surface – a technique that enables the use of inflatable formwork.

Monocoque Structures

These derive their strength from sandwiching a framework between thin panels. Known also as stressed skin structures, and commonly used in the aircraft, automobile and shipbuilding industries, they rely upon the outer panels to stiffen the structures – the outer skins carry all or most of the torsional and bending stresses.

Form Finding

During the last century, both architects and engineers developed ways to design complex, double curving, compressive surfaces by experimenting with physical models – specifically, by employing a three-dimensional version of the catenary curve. A net or fabric is suspended from a set of points and then fixed in position using plaster and/or glue. This is then flipped over (mirrored horizontally) to create a thin, shell-like form. Due to their structural efficiency – the model in pure tension becomes one in pure compression – these forms are known as minimal surfaces.

1 Anticlastic and Synclastic geometry: any surface that is curved in opposite ways in two directions – positively along one principle axis and negatively along the other (i.e. saddle-shaped) – is known as anticlastic; surfaces that curve toward the same side in all directions are known as synclastic. Hence, the catenary curve can be rotated to form a (synclastic) catenoid and the parabolic curve to form a (synclastic) hyperbolic parabola. Structurally, both of these forms may be applied either as thin, compressive shell structures or as tensile, cable nets.

2 L'Oceanogràfic, Valencia, Spain, Felix Candela. Reinforced concrete shell; a series of hyperbolic paraboloids meet centrally as in a groin vault.

3 Ruled Surface – a synclastic surface can also be formed using straight or ruled lines. This is not only convenient for the draftsman but also aids in generating a structural grid.

4 The Sydney Opera House, Sydney, Australia, designed by Jørn Utzon and engineered by Ove Arup (see also page 166). The Opera House was designed as a series of reinforced-concrete shells.

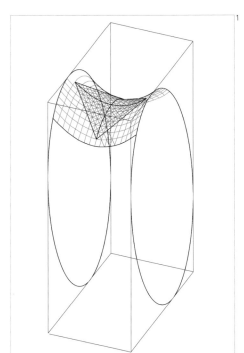

1

3

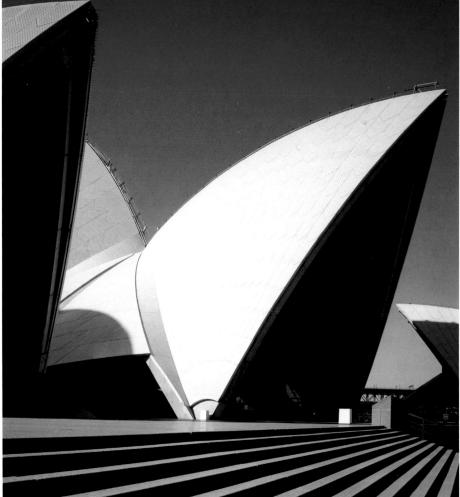

2

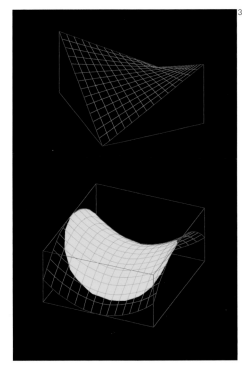

4

See also: The German Pavilion p160

1

Fabric

While some fabrics stretch to varying degrees, the design of pressurized or tensile membrane structures involves pattern-cutting techniques similar to tailoring: fabric seams are stitched and welded together, and may often be reinforced with cables (as in a cable net structure).

Air-Supported Membranes

In the case of airtight fabrics, membrane structures can be formed using air pressure alone. Superpressure (also known as inflatable or pneumatic) buildings maintain their shape by pumping air into the structure. As with a tire, air acts in compression to carry the loads, while the fabric maintains the overall shape under tension.

Cable Net

While a membrane structure relies partially upon its own elasticity to find its natural form, a cable net structure finds its optimum form under tension by having adjustable nodes that allow the cables to slide over each other wherever they interconnect.

Tensegrity

The word comes from tensile integrity, and Buckminster Fuller described the principle as "structure using distributed tension to hold islands of compression." Tensegrity produces minimal, lightweight structures wherein cables are tensioned against compression elements that are independent of (do not touch) one another.

1 Three-dimensional tensile surfaces. From left to right: mast-supported membrane, air-supported membrane, cable net, tensegrity.
2 North African Bedouin tents made from wooden posts and goatskin membrane.
3 German Pavilion, 1967 Montreal Expo, Canada, Frei Otto (see also page 160).
4 Mast-supported membrane structure. Anticlastic (saddle-shaped) geometry is inherent to efficient membrane structures.
5 Detail of membrane connection to mast.
6, 7 Millennium Dome, London, UK, Richard Rogers. A part cable net and part membrane structure, where the cables act as reinforcements for the tensile fabric.
8 Arch-supported membranes.
9, 10 Tennis court enclosure, London, UK, Birds, Portchmouth, Russum. Superpressure buildings maintain their shape by pumping air into the structure.
11 Tensegrity model.
12 Aviary, London Zoo, UK, designer Lord Snowdon, architect Cedric Price and engineer Frank Newby. A tensegrity structure.
13 Tensegrity sculpture by Kenneth Snelson. He describes the principle as "floating compression."

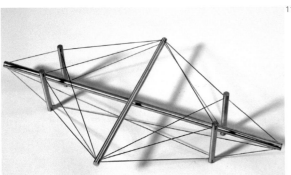

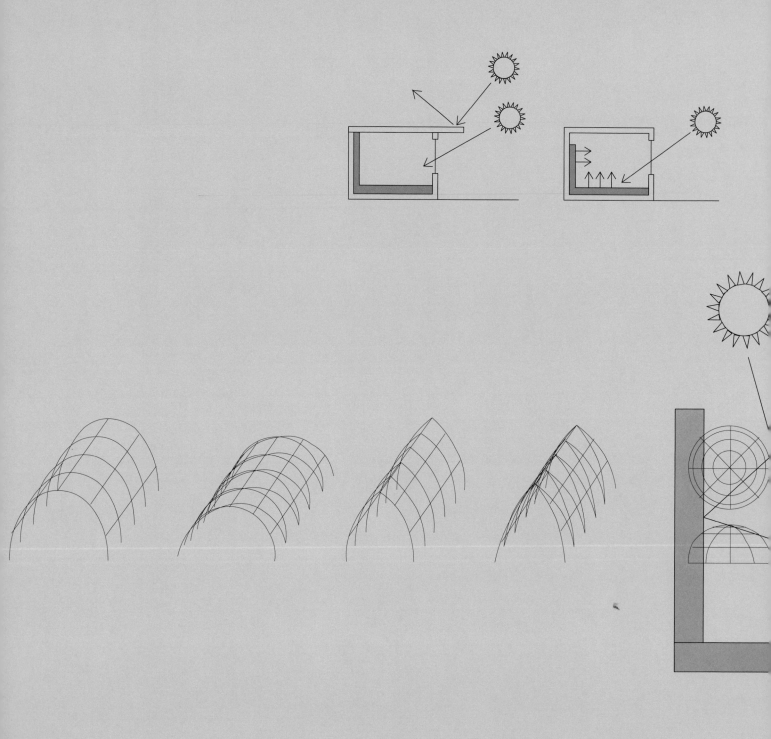

Climate and Shelter

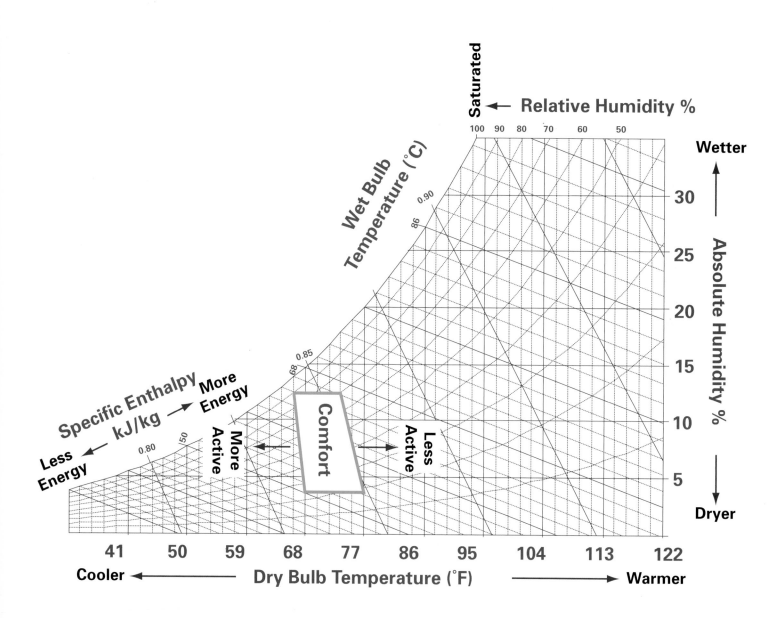

73 Human Comfort

Generally speaking, in architecture human comfort is dependent upon a number of factors:

Thermal Comfort
Maintaining the body at an acceptable temperature.

Air Quality
Ensuring the provision of clean air within enclosed spaces.

Illumination
Provision of adequate light levels.

Sound Quality
Clarity of communication and protection from noise pollution.

Sanitation
Providing for water distribution and waste disposal.

Subjective Perception and Adaptation
Human comfort is considered to be partially subjective, in that humans may perceive a room to be warmer or cooler than it actually is, e.g because of its color and materials. Variation in human comfort levels also results from evolutionary adaptations, i.e. where a body has adapted its specific physiology as a response to local environmental conditions.

Of these factors, however, it is a structure's ability to provide the body with good air quality that is the most important of its functions. Therefore, this section will focus on the principles of thermal comfort and its provider, the sun.

Left
A psychrometric chart is a graph of the thermodynamic parameters of (moist) air at a constant pressure; the chief parameters are wet and dry bulb temperature and absolute and relative humidity. The chart can therefore be used to map any site location (it may be adjusted for elevation relative to sea level) in order to map the specific, local needs of climate control in buildings. The yellow box encloses the area that humans consider to be "comfortable," however this box can be expanded in various ways by applying certain types of climate control. E.g., rather than heating or cooling the air inside a building for it to be within the normally accepted temperature range, air temperature can stray beyond this range if a radiant heating or cooling source is provided.

Thermal Comfort

Human comfort depends chiefly upon thermal comfort. Our core body temperature must remain at a constant 98.6°F. The body generates heat even while at rest. Indeed, the body must always be losing heat to maintain comfort because it produces more heat than it needs. In order to maintain equilibrium, the body must maintain a fairly steady rate of heat loss of around 300 Btu/hr when it is at rest and more when active. Lose too much heat and you feel cold, too little and you are hot.

Body heat is transferred through convection, conduction, radiation, and evaporation. When the ambient temperature is higher than skin temperature, the heat gained by radiation and conduction must be dispersed mainly through the evaporation of perspiration (see Passive Control/Evaporative Cooling, p. 88).

The body generates heat even while at rest. Indeed, the body must always be losing heat to maintain comfort because it produces more heat than it needs. When internal body temperature is insufficient the body starts to shiver, which in turn increases the production of body heat. In animals covered with fur or hair, "goose pimples" are created when muscles at the base of each hair contract and pull the hair erect as a response to cold. The erect hairs trap air to create a layer of insulation. As humans have lost most of their body hair, the reflex now serves little purpose.

The main factors influencing thermal comfort are:

Air Temperature

Air temperature is governed ultimately by solar radiation (see Solar Geometry, page 76).

Mean Radiant Temperature

Radiant temperature is governed both by the temperature of an object and its emissivity – its propensity to emit long-wave radiation. Mean radiant temperature (MRT) is the average radiant temperature of all objects within view of the subject and can vary significantly from air temperature (for example even a tightly sealed single pane window feels "drafty" on a cold winter day because its radiant temperature is much lower than other interior surfaces). Because we lose a substantial amount of heat via radiation, MRT is as important a determinant of comfort as air temperature.

Air Movement

Air movement (a breeze or draught) is governed by air pressure. A breeze of around 20 in per second provides an equivalent temperature reduction of around 5.4°F.

Humidity

High humidity levels reduce evaporation rates. For human comfort, relative humidity should be between 40 per cent and 70 per cent. However, when relative humidity exceeds 60 per cent, our ability to cool is greatly reduced.*

*Relative humidity is an indication of the water content of air. It is measured as the percentage of the actual water vapour density to the saturation vapour density (both are measured as mass per unit volume). Saturation vapour density is the amount of water vapour needed to saturate the air, and varies according to temperature. For example, if the actual vapour density is $\frac{1}{100}$ oz/ft^3 at 68°F compared to the saturation vapour density (at that temperature) of $\frac{1}{400}$ oz/ft^3, then the relative humidity is 57.1 per cent.

1 "Goose pimples."
2 A thermal image of the faculty of the thermal physics laboratory, Vanderbilt University, Nashville, USA.

76 Solar Geometry

Earth's Orbit

The earth orbits the sun in a counter-clockwise elliptical orbit once every 365.26 days. It spins counter-clockwise on its north–south axis once every day. (This accounts for the fact that the sun rises in the east and sets in the west.) This axis is tilted with respect to the plane of its orbit at an angle of about 23.4 degrees. The average distance from the earth to the sun is around 93 miles. In relative terms, if the earth was 1 inch in diameter, then the sun would be an 8 ft diameter disc around 984 ft away.

The Equinoxes

The equatorial plane divides the earth into halves – the northern and southern hemispheres. The intersection of the equatorial and ecliptic planes is called the line of equinoxes. One half of this line is the vernal (spring) equinox and the other half the autumnal equinox. At two points in the earth's orbit this line intersects the sun, making the start of the fall or spring season. Perpendicular to the line of equinoxes is a line which contains the solstices. These are the points that start summer or winter when they cross the sun.

The Calculation of Solar Radiation

To calculate the position of the sun on any given day at a certain place on earth, two angles must be specified: the solar altitude and the solar azimuth. The altitude angle is the angle in a vertical plane between the sun's rays and the horizontal projection of the sun's rays. The azimuth angle is the angle on the horizontal plane measured from the north or south to the horizontal projection of the sun's rays.

If we take a given location, e.g. Blackpool, UK, its location can be described as a pair of coordinates:

Latitude 53° 46' 12'' N
Longitude 03° 01' 48'' W

Using a sun-path diagram we can calculate that there is a maximum difference in altitude (how high the sun appears in the sky) between the winter and summer solstices of approximately 45 degrees and a maximum difference in azimuth (where on the horizontal plane the sun appears and disappears) between the summer solstice sunrise and sunset of approximately 270 degrees.

Summer Solstice Sun Position:
21 June 12:00 noon
altitude 59.73°, azimuth 180° (due south)

Winter Solstice Sun Position:
21 December 12:00 noon
altitude 12.73°, azimuth 180° (due south)

The hours of available (or potential) sunlight at the summer and winter solstices differ significantly. During the longest day (21 June) in Blackpool there is a potential 17 hours 6 minutes of sunlight from sunrise to sunset, whereas on the shortest day (21 December)

there are a maximum of 7 hours 24 minutes.

Location-specific sun path-diagrams afford the designer a basic knowledge of where and when direct sunlight will fall on any given design. It is worth remembering that sunlight can be described graphically as parallel rays, due to the sheer size of the sun in relation to the earth. It may also be useful to note that moonlight is reflected sunlight.

For more detailed analysis of solar geometry, software packages such as ecotect can be used to model a specific sunlight condition in relation to a particular design. See Environmental Analysis, page 128.

The WMO (World Meteorological Organization) states the definition of normal sunlight as 120W/m² or more shining on the earth's surface. Sensors such as pyranometers are available to measure this threshold.

1 Sun-path diagram for Blackpool, UK, showing solar altitude and solar azimuth in relation to time and date. http://solardat.uoregon.edu/SunChartProgram.html.
2 Altitude and azimuth.
3 Multiple exposure photograph showing the "midnight sun" visible in high northern latitudes.

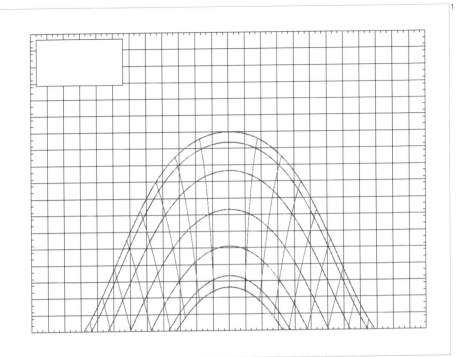

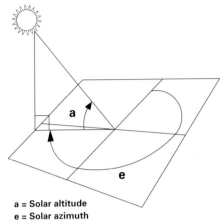

a = Solar altitude
e = Solar azimuth

HOT DAYS, COLD NIGHTS

Insulation prevents energy loss during daytime

By day, thermal mass acts as a heat sink and stores heat energy

During the day, insulation stops transfer of cold air from outside

GREENHOUSE EFFECT

Glass enclosure to trap solar radiation

DAY

At night, thermal mass reradiates heat

At night insulation prevents heat loss from inside

At night, insulation is required to prevent loss of heat energy to the outside

NIGHT

COLD DAYS, COLD NIGHTS

Insulation prevents transfer of cold air from outside

GREENHOUSE EFFECT

Low-emissivity glass reduces heat loss

DAY

At night, insulation prevents heat loss from inside

NIGHT

HOT DAYS, HOT NIGHTS

Insulation prevents transfer of warm air from outside

Cool by encouraging air movement through natural ventilation

Shading prevents solar gain during the hottest part of the day

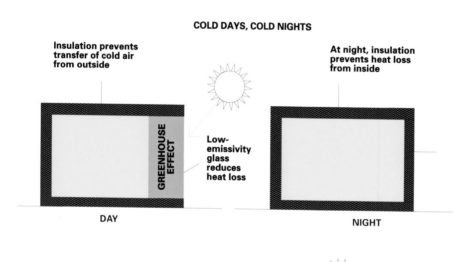

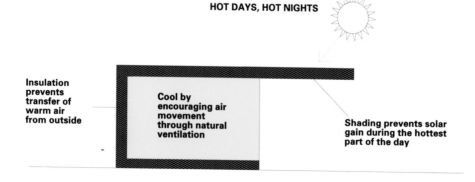

79 Building Performance

The Climatic Envelope

It is possible to design buildings to control their internal environments both actively and passively. A passively designed building "dances" with the ambient energy flows around it – a strategic arrangement of the form and fabric of the building captures and stores heat when it is needed, and avoids or releases it when it is not. An active environmental control strategy involves supplying or removing energy from the space through mechanical means such as solar-thermal hydronic heating, air conditioning, or a gas furnace. Buildings should be optimized to employ passive strategies and to rely on active strategies as required (using renewable energy whenever possible) to back up those passive systems. Even in cool climates, modern (well-insulated, airtight) offices tend to produce a level of heat from their occupants, lighting, computers, etc. so that they are more likely to need cooling than heating. With domestic environments, it is usually the opposite, with heating and cooling loads being dominated by outside conditions.

Passive Thermal Envelope

The form and orientation of a building dictate both solar gain, "the greenhouse effect," and natural air movement – air exchange. The fabric of the building dictates its ability to absorb energy, through thermal mass, and to present a barrier to energy flows, through insulation. Often, the architect has to deal with conflicting variables – in a hot climate a large expanse of glass offering a panoramic view, but facing the sun, will require shading during the hottest part of the day.

Sound and Light

Both sound and light are described as wavelengths which are measured as frequencies. The design of a building and the building fabric influence the ways in which both sound and light waves will behave. Acoustically, buildings have to maintain clarity of communication for their occupants and must protect them from noise pollution. Similarly, the maximum use of available daylight can contribute to both the well-being of a building's occupants and to energy conservation.

Mechanical Heating, Ventilation, and Air-Conditioning

While a certain amount of control over the energy used to heat, cool, and illuminate buildings can be achieved passively through the building fabric (design and orientation), it may also be necessary to employ additional energy sources and energy distribution systems. This section covers mechanical heating, mechanical ventilation and air-conditioning, although these may be designed in tandem with the passive envelope, as with hybrid active/passive cooling systems.

Left
Climatic envelope: basic principles of thermal comfort as related to the building envelope under different climatic conditions.

Thermal mass is the ability of a material to absorb heat energy. A good deal of heat energy is required to change the temperature of most high-density materials like concrete, brick, and stone. They are therefore said to have high thermal mass. Lightweight materials such as timber generally have low thermal mass. The specific heat capacitance is the property of a material that refers to the amount of heat required to change the temperature of a given mass of that material a certain amount. Water has a very high specific heat (1 Btu/lb °F) and relatively high density, giving it a tremendous ability to store heat with a minimal temperature change. Drums of water have even been used for thermal mass walls in some cases.

Effective use of thermal mass in buildings will tend to moderate temperature extremes, with the resulting indoor temperature oscillating around the average outdoor temperature in the absence of a heat source (such as passive solar gain) or a heat sink. In addition to the reduction of temperature fluctuation, there will also tend to be a lag between outdoor conditions and the indoor temperature response. For example, in an environment subject to hot days and cold nights, thermal mass may be used to absorb heat during the day from direct sunlight (we experience this when walking into a cathedral, mosque, or castle in a hot climate) and it will re-radiate this warmth back into the buidling during the night. Thermal mass acts as a "thermal battery" by storing excess heat and releasing it during cooler periods.

Strategic use of thermal mass is an essential and time-tested passive strategy. Most modern high-mass materials also have a high thermal conductivity (e.g. concrete) and therefore need to be placed inside of insulation so that their stored heat is not released to the outside before it becomes useful in the building, and to minimize excessive heat flow through the building envelope.

Trombe Wall
The trombe wall is a specialized application of thermal mass. Heat from the sun is trapped between window glazing and a dense, masonry wall (see Windows and Glazing, page 90). The wall stores heat for release into the building later in the day. Edward Morse patented the design in 1881. However, it really became efficient only with the advent of insulated glass (to prevent heat loss back to the outside) and with the introduction of either natural or fan-assisted ventilation to encourage heated air to flow into the interior through convection. The space between the window and the wall then becomes a solar thermal collector.

1 Direct solar gain.
2 Indirect solar gain: trombe wall.
3 Labyrinth, Federation Square.
4 Federation Square, Melbourne, Australia, Lab Architecture Studio with Atelier 10. This "thermal labyrinth" acts as a battery pack that can be flushed with colder night air during summer so that it can cool incoming air during the day. The basic idea is simply that air is exposed to a large surface area of concrete on its way into the building by passing at low velocity down a tunnel.

3

Direct solar gain

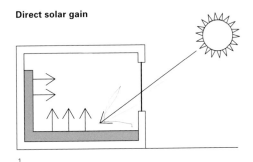

1

**Indirect solar gain
(Trombe wall)**

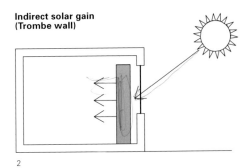

2

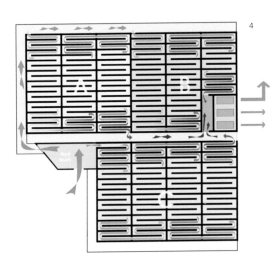

4

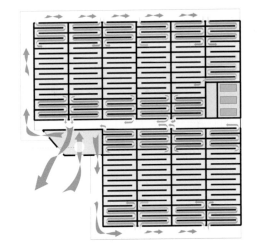

Insulation acts as a barrier to heat flow and is essential to keep buildings warm or cool. Depending on the climate, it is necessary to establish whether the insulation is predominantly needed to keep heat in or out (or both).

Conduction, Convection, and Radiation

Heat will flow (or transfer its energy) by one of the following mechanisms. Conduction is the transfer of energy through matter from particle to particle, from atom to atom. For example, a spoon placed in a cup of hot liquid becomes warmer because the heat from the liquid is conducted along the spoon. Convection is the transfer of heat energy in a gas or liquid by movement of currents: heat leaves the hot cup as the currents of steam and air rise. Radiation is the direct transfer of energy, without the aid of fluids or solids, through electromagnetic waves; sunlight radiates heat to our planet.

Thermal Conductivity

Materials with a high thermal mass are not necessarily good insulators. Good insulative properties depend upon the ability of a material to resist heat passing through it. Such materials are said to have poor thermal conductivity. For example, rubber is a poor conductor of heat, brick is good, concrete is better. The surface of a material will affect its thermal conductivity through its ability to reflect heat energy. Dark, matt or textured surfaces absorb and reradiate more energy than light, smooth, reflective surfaces.

U Value

The ability of a material (such as in a window, wall, or roof) to transfer heat energy from one side to the other is measured as the thermal transmission rate per area per temperature difference. It is referred to as conductance or U value. The formula is expressed as Btu per hour per square foot of surface per degree Fahrenheit ($Btu/hr/ft^2 \, °F$), and the resulting figure is known as the U value or conductance.

R Value

The insulative qualities of walls, floors, and roofs are generally reported in terms of resistance or R value. This is simply the arithmetic inverse of U value and is expressed as ($hr \, ft^2 \, °F/Btu$). Resistance values are useful because the R value of the individual layers of an assembly may be totaled to obtain the value for the entire section.

Note: A penguin is able to maintain an internal body temperature of around 95°F when the outside temperature is -58°F. A ¼ in layer of down (tiny feathers measured in microns) produces millions of tiny air pockets to insulate the penguin.

1 Where thermal mass is employed to store heat energy, an external layer of insulation prevents heat reradiating to the outside. Similarly, insulated glazing can prevent heat loss from a Trombe wall at night.
2 Thermal imaging techniques (thermography) use infrared cameras to identify heat flow through buildings.

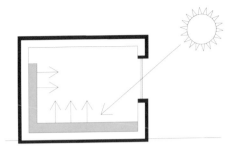

1

2

Products used to enhance the insulative capacity of buildings currently use two basic methods, bulk insulation or reflective insulation.

Bulk Insulation

This resists the transfer of conducted or convected heat by relying on pockets of trapped air within its structure. Its thermal resistance is essentially the same regardless of the direction of heat flow through it. Bulk insulation includes natural materials such as cotton, wool, and cellulose, as well as synthetic materials such as fiberglass batts, polystyrene, and polyurethane foams.

Foam insulations have the added benefit that they seal the building envelope, limiting air leakage.

Radiant Barriers

Radiant barriers usually have an aluminum foil surface and must be separated from adjacent materials by an airspace so that heat cannot be conducted directly through the material. The radiant barrier then becomes a very effective defense against heat transfer because it has a very poor ability to absorb or re-radiate long-wave radiation. This property is called low emissivity. Because the surface is not in contact with other materials, it cannot conduct heat to them directly, so the only heat transfer pathway left is convection.

Thermal Bridging

This is a term used for situations where a thermally conductive material traverses from the inside to the outside of the building, allowing heat or cold to pass through it.

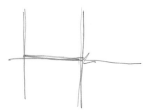

1 Bulk insulation in a cavity wall. This type of insulation may also consist of a thinner, multilayered fabric combining bulk and reflective insulators. Other useful insulators include paper pulp, sheep's wool and penguins (well, not penguins actually).

2 Aerogel: created by Steven Kistler in 1931, Aerogel is a low-density material in the form of a gas-filled gel. Aerogels made from silica and carbon are able to nearly eliminate heat transfer by convection and conduction.

3 Sections through walls of differing composition, both producing a 53° F difference between the inside and outside.

cold

1

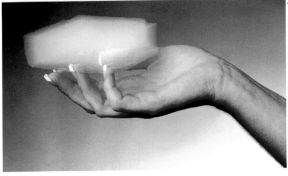

2

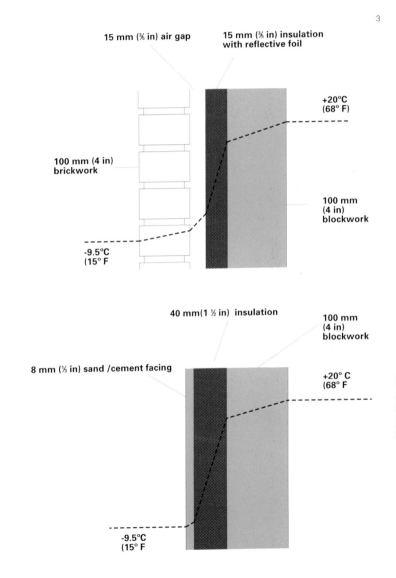

3

15 mm (⅝ in) air gap

15 mm (⅝ in) insulation
with reflective foil

100 mm (4 in)
brickwork

+20°C
(68° F)

100 mm
(4 in)
blockwork

-9.5°C
(15° F

40 mm(1 ½ in) insulation

100 mm
(4 in)
blockwork

8 mm (⅛ in) sand /cement facing

+20° C
(68° F

-9.5°C
(15° F

Ventilation is the exchange of stale air with fresh air within a building. This can have a number of effects. It can supply clean air, i.e. dilute carbon dioxide and supply oxygen; it can replace hot air with cool air; and it can lower relative humidity and prevent condensation (where water vapor in the air collects on surfaces). Ventilation may also be needed to remove smoke in the event of a fire.

Rate of Change

Ventilation is normally measured in terms of the number of times that the air in a room is completely replaced in an hour (air changes per hour: ac/h). For a typical office the recommended rate is between 4 and 6 ac/h; for a restaurant, it is between 10 and 15 ac/h.

Natural Ventilation

Naturally ventilated buildings are designed to promote air exchange without the use of mechanical means. Natural ventilation relies on pressure differences to move air from areas of higher pressure to areas of lower pressure. When windows are open on two sides of a space, and a breeze is coming from one direction, positive pressure is produced on the windward side and negative pressure on the leeward side of the building. This pressure differential tends to suck air through the space. Cross ventilation is the most basic of natural ventilation strategies.

Stack Effect

The stack effect is a form of convection. Because warm air is less dense than cold air, a building with openings at both the top and the bottom will experience a pressure difference. In the hot season, the warmer indoor air will rise (through thermal buoyancy) and reduce the pressure at the base, encouraging denser, cooler air to infiltrate. The greater the thermal difference and the height of the structure, the greater this effect. In a tall building with a sealed envelope the stack effect can create significant pressure differences: stairwells and lift shafts, etc. need to be controlled to prevent the spread of smoke in the event of a fire.

A chimney will encourage the stack effect by acting like a venturi nozzle – where gas from a large chamber accelerates when passing through a small chamber into a second large chamber due to the pressure differences (similar to the uplift developed by an airfoil shape which creates a pressure differential above and below a wing). A chimney will draw air vertically through a wind or breeze passing over its top and creating a pressure difference.

1 Cross-ventilation. Warm air rises or, more accurately, heat flows from areas of higher temperature to areas of lower temperature.

2 Natural air flow assists convection of warm air from a Trombe wall into the interior.

3 Wind is described in terms of the direction from which it originates, as read on a compass. For any location, wind will, on average, tend to come from a particular direction; this is known as the prevailing wind.

4 A wind map, known as a wind rose, is used to describe the average wind direction for any location, as in this random example. Each circle represents the percentage of time the wind blows from a particular direction. Such information can be used when orientating structures to make use of natural air movement or to assess wind impact on a façade.

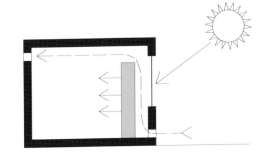

1

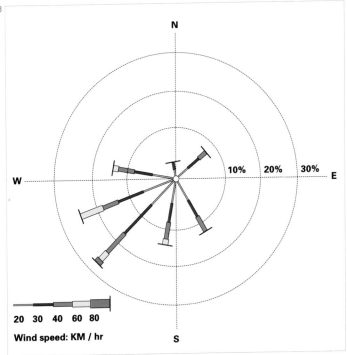

2

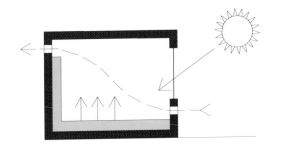

3

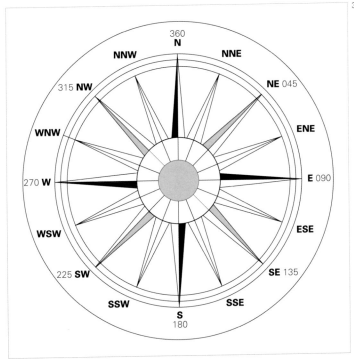

4

Climatic Envelope / Air Movement: Passive Control

See also: Tepee p136
30 St. Mary Axe p176
Davies Alpine House p180

In temperate climates, adequate ventilation can usually be achieved by small openings – trickle ventilators – that are positioned to maximize air movement and wind pressures from the local environment. These can be controlled – opened and closed – using manual "dampers." The building envelope should be as airtight as possible, with fresh air introduced intentionally as needed. For rapid ventilation of rooms, windows can be made to open!

Wind Scoops and Wind Cowls

While a wind scoop employs the principle of catching the wind and forcing it downward into the building (and is therefore constructed to offer an opening towards the prevailing wind), a cowl such as those used on traditional oast houses works in the opposite way to encourage air to be sucked out of the building.

Solar Chimney

A centuries-old idea, the solar chimney is a way of amplifying the stack effect by heating the chimney to produce greater temperature and pressure differences. Chimneys are constructed with a large area of thermal mass (which may be painted black) facing the sun. An updraft is created by the added buoyancy of the air in the chimney, creating suction at openings at the building's base which ventilate the space.

Evaporative Cooling

If you wet the back of your hand and then blow on it, your skin surface feels cooler. That's evaporative cooling. When air comes into contact with water, some of the water is evaporated into the air unless it is already at saturation. In the phase change process from liquid water to vapor, the air transfers some of its sensible heat (heat that affects air temperature) to latent heat (energy used to achieve the phase change).) The result is that the air temperature drops and the relative humidity rises; a very effective comfort strategy in hot-dry climates, but not in hot-humid ones. The amount of temperature drop is largely a function of the amount of moisture already present in the air. Evaporative cooling is a time-honored tradition in many desert cultures; for example the Islamic courtyard pool provides both a source of evaporative cooling and a psychological cooling effect.

Passive Down-draft Cooltower

Another hot-dry climate cooling technique is the passive down-draft cooltower. Water is pumped over pads located at the top of a hollow tower. This water evaporates, cooling the air. The cool, dense air "falls" through the tower and exits through large openings at its base. Modern mechanical evaporative coolers work in much the same way.

1 Passive down-draft cooler.
2 The principle employed by a wind cowl is that air passing around the curved surface creates a negative pressure around the opening, thus sucking out the air from within.
3 Traditional Middle Eastern "wind catchers." A wind catcher is a tower that is capped but has openings along the sides. By closing all but the one side that faces away from the incoming wind, air is drawn upwards; due to the height of the tower, the pressure gradient that results from a passing breeze helps to suck the warmer air out from below. Opening the side facing the wind would push air down the shaft.
4 Traditional Middle Eastern wind scoops.
5 Oast houses were used in southern English counties for drying hops. Cowls on the roofs provided an exhaust for the hot air used in the drying process and were able to rotate by using a wind vane.
6 Modern wind cowls at Beddington, UK, are designed to maximize the effect of a passing breeze, both through aerodynamics and by swivelling as the wind changes direction.
7 Passive down-draft cooltowers.

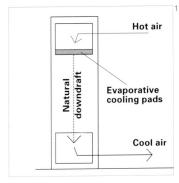

Hot air

Evaporative
cooling pads

Natural
downdraft

Cool air

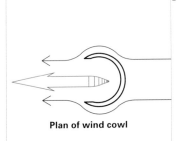

Plan of wind cowl

90 Climatic Envelope / Windows and Glazing

See also: Tropical House p152
The USA Pavilion p162
HSBC Headquarters p168

Windows typically represent the most important pathways for heat gains and losses in modern buildings. In hot periods the transmission of solar radiation into a building through its windows can cause overheating, while in cooler climates heat loss from inside the building through windows can greatly reduce the building's warmth. These factors can be mediated through solar shading and through the type of glazing used.

Balancing Gains and Losses

As with all heat transfer, heat will flow from the warm side to the cool side of a window. Windows are usually poor insulators compared to solid walls and roofs, and they are nearly transparent to solar radiation, so they represent a large component of the heat gains and losses of a building. Windows facing the sun often gain more solar energy during the day than they lose at night. There are several processes that influence rates of heat loss through glass and window components. Heat loss through windows takes place through both the glazing and frame edges mainly by conduction.

Double Glazing

The insulative quality of a window can be greatly improved by sandwiching two or more panes of glass together using spacers along the edges to produce a sealed air gap. The best spacing to minimize convection losses in double glazing is estimated at ½-⅝ in; stationary air is not a good conductor. Window frames must also be designed in

such a way to maximize insulation. Thermally speaking, preferred frame materials are those with low conductivity (such as wood or fiberglass), or frames that are thermally broken, such as aluminum with a low-conductivity plastic strip that connects the inner and outer portions of the frame section.

The Greenhouse Effect

Solar radiation reaching the surface of the earth has a short wavelength (because the surface of the sun is hot – 6000 K). However, window glass is better at blocking the long-wave radiation that is re-emitted from internal surfaces (300 K). Short-wave visible light is converted to heat when it is absorbed by materials inside a space, then re-emitted as infrared radiation that cannot pass directly back through glass. Therefore, temperatures increase quickly as heat builds up. This phenomenon is known as the greenhouse effect.

Emissivity

Because ordinary glass readily radiates heat, i.e. has high emissivity, its ability to retain heat can be improved by lowering its emissivity – hence the term low-emissivity or low-E glass. This is achieved with a coating of microscopic pieces of metal on one face of the glass. Some coatings help to reject incoming solar radiation (reduce solar gain).

Other Coatings

Spectrally selective coatings serve to reduce heat gain by blocking radiation of those

portions of the electromagnetic spectrum that are not visible to the human eye. Tinted coating limit certain wavelengths of visible light)as well as invisible radiation in some cases) and reduce overall light transmission. Mirrored coatings also cut down on visible light transmission into the building. Coatings that cut visible light transmission significantly, however, may result in additional lighting requirements which reduces their energy-saving benefits as well as denying occupants the full benefits of daylight access.

1 The greenhouse effect.
2 Roof overhang designed according to seasonal changes in solar geometry.
3 Brise-soleils, also known as louvred blinds, are designed to prevent solar ingress while maintaining daylight and views.
4 If louvres are placed between glass panels with vents at the top and base they will absorb solar radiation and the heat will be removed by natural convection.
5 Deciduous trees allow the sun to warm a building in winter while shading it in summer.
6 Adjustable brise-soleils – horizontal for predominantly high sunlight.
7 Adjustable brises-soleils – vertical for predominantly low sunlight.
8 Types of double-glazing.

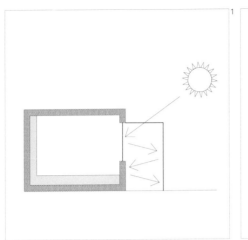

1

2

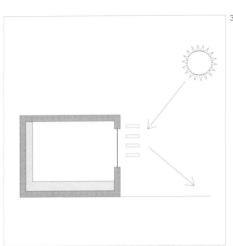

3

4

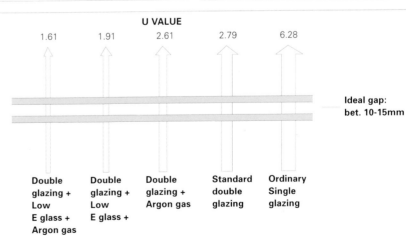

5

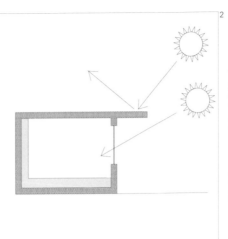

6

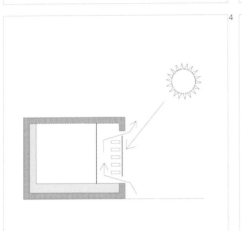

7

8

U VALUE

1.61	1.91	2.61	2.79	6.28

Ideal gap:
bet. 10-15mm

Double glazing + Low E glass + Argon gas

Double glazing + Low E glass +

Double glazing + Argon gas

Standard double glazing

Ordinary Single glazing

92 Climatic Envelope / Weathering: Masonry

Whether used for loadbearing, or as infill/cladding for a framed structure, a wall built from stone or concrete blocks or from clay bricks will resist water penetration and wind load if properly constructed. In walls constructed with mortar beds, rain is most likely to penetrate through the joints or where the masonry joins to another material such as at a window opening. Brick, stone or concrete blocks are often pointed along the joints – a thin (roughly ⅜ in) layer of a high-density mortar is applied in order to prevent water penetration. Where movement is likely, due to differential thermal expansion and contraction of materials (such as at door and window openings), a flexible – plastic – material is required, which must also be able to adhere to more than one type of material: examples are lime, putty, and synthetic mastic.

Face Brick

Face brick is the most common type of brick and today is made to standards that require the brick to comply with low water absorption standards. Although a well-built brick wall is water resistant, some moisture inevitably makes its way through the exterior of the envelope. Masonry systems either act as a reservoir to store and gradually release infiltrating moisture (solid masonry construction), or employ a cavity wall system to drain and release any water that makes it past the surface veneer. Two major types of defect can occur in brick construction due to water. Spalling occurs in soft sub-standard brick when absorbed water freezes and crumbles or splits off layers of the brick face. Efflorescence is a deposition of dissolved salts on the surface of the brick that usually results from rainwater leakage into the system where the moisture evaporates back out through the face veneer.

Concrete

Concrete is water resistant, although this varies according to the precise mix and construction techniques. In cast reinforced-concrete walls or columns, steel reinforcing bars may be subject to corrosion if placed too close to the surface of the concrete.

Tiles

Clay, stone, and concrete, as well as other artificial materials, are used to weatherproof roofs and walls in the form of small modular panels – tiles. Strictly a form of cladding (see Cladding Systems, page 94), this traditional method of waterproofing has the advantage that the small modules are easily handled and replaced if they are worn or damaged. Tiles are hung from a subframe in a lapped and/or interlocking manner.

1 Pointing: the mortar joints between brick courses may be detailed in various ways on the outer face. Often a thin layer of harder mortar will be applied to the surface. **A** Flush, **B** Recessed (raked), **C** Bucket Handle, **D** Weatherstruck.

2 Weatherstruck pointing: a ratio of around 2 parts fine sharp sand to 1 part cement.

3 Cast-concrete wall. The finished surface remains as it was when the timber shuttering was removed, showing the imprint of the wood grain.

4 Clay roof tiles have been profiled to interlock with one another and improve rainwater run-off. These are known as pantiles.

5 Slate tiles have been nailed to a subframe of horizontal timber batons to form a vertical cladding.

94 Climatic Envelope / Weathering: Cladding Systems

With framed construction and where masonry is not used as infill, it is usual that a subframe is constructed in order to carry the weatherproof "skin" of the building. This sub-frame, along with the skin itself, must also be designed to resist wind load. Subframes may be constructed as infill panels, as curtain walls, or as independent structures ("outriggers") that carry the entire load of the outer skin.

Cladding Panels

There are many materials that can be manufactured in sheet form and that are either naturally waterproof or may be treated or coated to provide water resistance. Metals such as lead, zinc, steel, and copper may be used, as well as glass-reinforced epoxy (GRE), polycarbonate and, of course, glass. Cladding panels may be composed of layers of different materials (sandwiched or laminated) and opaque panels will usually include some type of lightweight (e.g. foam) core for insulation and stiffening. Panels are normally housed in lightweight subframes (typically aluminum) that are then attached to the superstructure. Flexible connections and sealants are needed for differential thermal expansion.

Ventilation

For the purposes of ventilation, louvres and operable windows or vents are often incorporated into cladding systems.

Curtain Walls

Curtain walls are a type of cladding that is hung off the edge of the floor plate. They transfer their loads to the main structure through independent connections to floors and/or columns.

Planar Systems

By employing special fixings (sometimes known as spiders), glass panels can be attached to a subframe in such a way that they form a continuous surface. The panels are sealed at their edges using a flexible mastic.

Rain Screens

Rain screens are an advanced method of rainwater management that are beneficial in any location, but are particularly desirable in areas with heavy rainfall and wind. The rain screen employs a cladding surface in front of a well-ventilated cavity that separates the water resistant drainage plane from the surface veneer. The rain screen deters rainwater from entering the cavity and equalizes pressure on both sides of the cladding to reduce the chance of moisture penetration beyond the drainage plane. Moisture that does enter the cavity is easily drained out or evaporated via ample ventilation. The vented cavity may provide thermal benefits to the building through self-shading and heat removal from the vented cavity in some cases. Many different cladding systems may be employed in a rain screen system.

1 Lightweight (cold-rolled steel) metal stud framework used to infill main concrete structure.
2 Lightweight, hollow clay bricks used as infill panels.
3 Sheet materials can be formed into stiff panels (e.g. through corrugation or profiling) and can then be fixed directly to the main structure of the building.
4 Wood cladding. Wood sections may be used for weathering and are either horizontally lapped – "shiplapped", as in a boat hull – or interlocked through "tongues and grooves" cut along the planks. Planks are attached to a subframe, which is in turn fixed to the main structure. Timber starts to decay if its moisture content is much beyond 20 per cent so any exposed timber is normally coated with waterproof paints or varnishes.
5 The surface of copper sheet cladding oxidizes and turns green over time if exposed to moisture, a process known as patination. This does not effect its resistance to water penetration.
6, 7 Cladding panels are prefabricated with built-in brackets for fixing to the main framework.
8 A planar glass screen is fixed to a lightweight steel subframe employing vertical trusses.
9 Glass curtain wall.

Climatic Envelope / Weathering: Membranes

Membranes are waterproof fabrics that can be applied in a variety of ways. In tensile structures, the membrane acts to keep out the wind and rain. While tents were originally made from animal skins, ripstop nylon is the modern material from which tents are made. A lightweight, water-repellent fabric, it is reinforced with threads interwoven in a cross-hatch pattern so the material resists ripping or tearing.

PVC

Polyvinyl chloride is a thermoplastic, a by-product of the petrochemical industry, and is produced in the form of a white powder that is blended with other ingredients to form a range of synthetic products. Membrane fabrics are made from PVC-coated polyester. These fabrics are strong in tension and resistant to shear. They are light, flameproof and have a maximum thickness of about ⅟₁₆ in. There is significant cause for concern related to the environmental impacts of PVC. The manufacture of this plastic releases highly toxic dioxin into the atmosphere and dioxin is once again released at the end of its life cycle if the plastic is burned. Additives with adverse human health impacts in the form of phthalates and heavy metals among others are used to plasticize and stabilize PVC for many applications.

ETFE

Ethylene tetrafluoro ethylene is a fluorocarbon-based polymer. It was designed to be a material with high corrosion resistance and strength over a wide temperature range. Thin and lightweight, it is also extremely translucent.

PTFE

Polytetrafluoroethylene is a fluoropolymer that has the lowest coefficient of friction of any known solid material.

Breather Membranes

A breather membrane is a woven material with spacings large enough to allow water vapor molecules to pass through from inside to outside yet small enough to restrict water liquid molecules coming the other way (microporous). It is therefore both waterproof and vapor permeable and is commonly used as a secondary barrier (underlay) below tiles and slates to aid ventilation within pitched roofs. Breather membranes are also known as vapor permeable underlays (VPUs), and are made from heat-laminated spunbond polypropylene with a polyethylene film.

1, 2 Tensile, waterproof membrane made from PVC.
3 The Aquatic Centre, Beijing, China. The structure is clad with hexagonal, inflated "pillows" – double-skinned, pneumatic panels – made from ETFE.
4 Detail of ETFE "pillow" air supply.
5 Tyvek is a brand of spunbonded high-density polythene made by DuPont. It is used as a contemporary replacement for traditional asphalt-based building papers. Such "breather" membranes prevent rain ingress, while allowing moisture to pass out of the building interior.

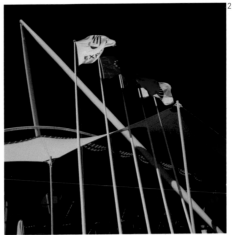

Felt

Roofing felt is composed of a polyester or glassfiber base saturated and coated with oxidized bitumen and surfaced with either bonded or loose sand or gravel.* Fixed onto a decking with cold applied adhesives or hot bonding bitumen, it is predominantly used as a built-up two- or three-layer system. A final surfacing is generally applied in order to resist general wear and for UV protection; this can be asphalt,** a granule-surfaced cap sheet or loose gravel. Some roofing products incorporate titanium dioxide in the top layer, specifically for UV resistance.

Ethylene Propylene Rubber

This is the basis for a number of products that consist of a single-ply roofing sheet that employs either welded or taped seams along the joints.

Stucco

When the quality of the substrate, be it brick, stone or concrete, is inadequate either from a weathering or aesthetic point of view, a layer of sand and cement mix known as stucco or render may be applied to the face of the building. The wet mix is usually built up in a series of layers for maximum weatherproofing. It may be applied directly to masonry or concrete walls. However, it is also used to weatherproof timber-framed structures where a damp-proof membrane (DPM) is fixed to the timber panelling, followed by metal mesh or timber laths to provide a key for the mortar. Stucco is often

painted and may be combined with other natural or artificial additives such as small pebbles (pebbledash) or tiles and mosaics.

*Known as a damp-proof course (DPC), similar products are incorporated into solid walls at low level to prevent "rising damp", where moisture is sucked up into the wall from the ground below.

**The word asphalt in British English refers to a mixture of mineral aggregate and bitumen (also known as tarmac).

Roofs

As a general rule, roofs will dispose of rainwater in one of two ways. They will either overhang the façade and eject rainwater away from the building (as in 5), or they will contain the water in the manner of a swimming pool (as in 2 and 3). The former are likely to be sloping or pitched to encourage run-off (usually in a controlled manner through gutters and downpipes), while the latter require a completely waterproof membrane and a surrounding dwarf wall or parapet with an integrated drainage system.

1 Over 1m concrete tiles are glued onto the shells that form the roof of the Sydney Opera House, Sydney, Australia.
2 Single-ply roofing sheets with "standing" seams.
3 Felt roof bonded to a timber platform with bitumen.
4 Rendered façade.
5 Pebbledash render.

100 Climatic Envelope / Weathering: Insulation

Cross-section detail of a glazed curtain wall

Curtain walls are a system of cladding a building in which the cladding elements are normally hung from the edge of the floor plates; the curtain wall façade does not carry any load from the building other than its own dead weight and transfers horizontal (wind) loads through connections at floor level. Here, an extruded aluminum frame is infilled with glass, although all types of cladding panel can be employed as curtain walling.

Cross section detail of ETFE air pillow

ETFE (ethylene tetraflouroethylene) is a man-made flouropolymer whose principle ingredient is fluorite, a common mineral. Extruded ETFE film was first developed in the 1970s, its use as an architectural cladding element being pioneered by Vector Foiltec. The film is deployed as pillows or cushions inflated by air. Fabricators have designed a variety of aluminum sections to grip the foil and create a framing system for connection to a super-structure. Shown here are also the flexible tube connections for maintaining air pressure within the pillows.

Cross section detail of a pitched roof

Pitched roof consisting of structural sawn timber sections and weathered using clay tiles. The apex is clad with a bespoke ridge tile and rainwater is collected in a half-round profiled gutter affixed to a timber facia board.

Cross section detail of sloping roof

Sloping roof shows insulation panels and a glazed opening. The junction of the upper end with the structural wall shows a detail for preventing water ingress while the lower end shows rainwater falling into a square-profiled gutter.

Cross section detail of brick-faced cavity wall

Vertical section cut through an external wall and printed at a scale of 1:20. At this scale it is possible to describe the shape (by line) and materiality (by fill) of the components that go to make up the wall. This is a brick and block cavity wall with an outer course of bricks, an air gap with a layer of insulation inside the cavity, and concrete blocks forming the inner leaf. A further layer of insulation and a facia panel (together known as dry-lining) are applied internally. The openable (casement) window is constructed from a timber frame with double-glazing, and the floor is made up of a layer of compacted hardcore, a waterproof membrane, a concrete pad, insulation, and a floor surface set on timber batons. There is also a concrete window sill – the little notches in the sills are known as drips and prevent water from seeping back into the brickwork by capillary action. The rebates in the timber frame are for molded seals (i.e. rubber gaskets). There is a reinforced concrete strip foundation.

1 Cross section detail of a glazed curtain wall
2 Cross section details of ETFE air pillow
3 Cross section detail of a pitched roof
4 Cross section detail of a sloping roof clad in lead
5 Cross section detail of brick-faced cavity wall

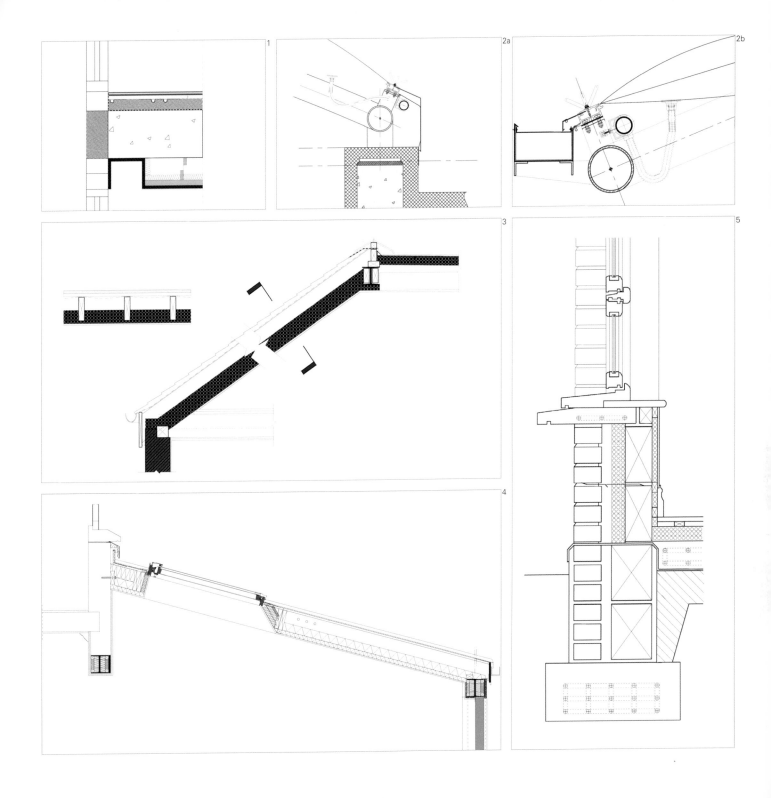

1

2a

2b

3

5

4

When the renowned Czech tennis player Ivan Lendl was asked the perennial question "what makes Wimbledon so great?", he answered without hesitation: "It is the sound of the balls." Acoustics is the science of sound. Sound or sound waves are a physical disturbance of molecules within a medium, gaseous, liquid or solid, that can be heard by the listener. This molecular disturbance comprises high and low pressure zones described as a wave cycle.

Frequency

The frequency of a sound is determined by the number of wave cycles per second (cps), which is more commonly expressed as Hz after the nineteenth-century German scientist Heinrich Hertz. Good human hearing can hear frequencies from as low as 20 Hz up to 20,000 Hz (20 kHz). Human speech has a more limited frequency range from 200 Hz to 5,000 Hz (5 kHz), whereas a concert orchestra can produce frequencies from 25 Hz to 13,000 Hz (13 kHz). Frequencies below 20 Hz (infrasonic) are sensed as vibration. Often we experience both sound and vibration within buildings. Sound travels at 330 m/sec in air, four and a half times faster in water and faster still through solid substrates. Sound does not travel in a vacuum.

Loudness

Sound pressure determines loudness, and the human ability to detect a huge range of pressure variation led to the use of the decibel (dB – a logarithmic scale) as the standardized measure of sound pressure level (SPL). The dB relates to the human perception of loudness with the threshold of our hearing represented by 0 dB and 140 dB representing the threshold of pain and potential hearing damage. From the quietest to the loudest sound we can endure is a factor of 1 to 10m; an added 20 dB marks a tenfold increase in SPL. Sounds are a combination of frequencies and loudness affected by time.

Reverberation

"You are listening to the room." Reverberation can be described as the build-up of multiple sounds arriving indirectly to the listener via surface reflections. The reverberation time (r/t) of a room is measured as the time taken for the sound-pressure level to fall by 60 dB. Reverberation time is related to the volume and geometry of a room and the absorption coefficients of the surfaces of that space. The reverberation time of any given space will affect the reproduction and intelligibility of a specific sound source. An optimum r/t for speech and electronically amplified music is 1 second, whereas an orchestra may require $1\frac{1}{2}$– $2\frac{1}{2}$ or more. The r/t of St Paul's Cathedral in London is 12 seconds. With a short r/t you can precisely pinpoint the sound source, which is useful during a lecture or speech, whereas an orchestral performance fills the room with sound (and reflected sound); it is the interaction between physical structure, space and music that you are listening to. In an electro-acoustic environment, where sound is picked up via microphones and then transmitted through a network of speakers, it is possible to shorten or lengthen the r/t, thus altering the notional "acoustic size" of the room and creating a "multi-purpose" space.

Typical Noise Levels

Typical Noise Levels	Loudness dB
Threshold of hearing	0
Whispering	20
Quiet street	40
Quiet conversation	50
Loud voice / busy office	60
Average traffic	70
Slamming door	80
Motorcycle	90
Nightclub	100
Pneumatic drill	110
Jet aircraft take-off (at 64 ft)	130
Threshold of pain	140

London Noise Map

Part of the Department for Environment Food and Rural Affairs (DEFRA) ambient noise strategy.

Loudness dB

| 80 |
| 75–80 |
| 70–75 |
| 65–70 |
| 60–65 |
| 55–60 |
| 50–55 |
| 45–50 |
| 40–45 |
| 35–40 |

Reverberation in a Building

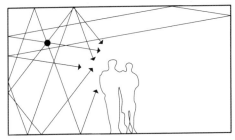

The table below shows typical reverberation times measured at 125 Hz, 500 hz and 2000 hz; these times are averaged out.

Typical Reverberation Time	r/t secs
Outdoors	0.0
Bedroom / Living room	0.4
Cinema (recommended)	1.0
Glyndebourne Opera House (new), UK	1.3
Royal Festival Hall, London, UK	1.4
Concert Hall (recommended)	1.5
Carnegie Hall, New York, USA	1.8
Church (recommended)	2.0
Musiksvereinsaal, Vienna, Austria	2.05
Symphony Hall, Boston, USA	2.2
Symphony Hall, Birmingham, UK	2.4
St Paul's Cathedral, London, UK	12.0

104 Sound and Light / Acoustics: Acoustic Control

Reflection

Reflected sound is the cause of reverberation, but you do not necessarily eliminate all unwanted reflections by controlling the reverberation time in a space. Acoustic reflections are a result of the geometry of a space and its material construction. Reflective parallel surfaces can cause standing waves in a room, which produce an unwanted "fluttering" effect. Domed and concave surfaces can focus reflections and cause "loud spots," and also direct sound around a space in the case of the Whispering Gallery in St Paul's Cathedral in London. Similarly, the designer can use the geometry of reflecting surfaces to help direct sound acoustically to all parts of a room in the case of a concert hall, theatre or other type of auditorium. The use of a parabolic surface behind a sound source will project a parallel beam of sound, with reflections on smooth, flat surfaces tending to conform to Lambert's Law: the angle of incidence equals the angle of reflection.

Absorption

The sound characteristics (acoustics) of a room depend on its size and the materials that clad and fill it. Porous materials and substrates such as carpets, curtains and mineral fibers deaden the oscillation of sound waves at high frequencies through friction, where sound-energy is converted into heat. Panel absorbers (often of timber construction) deaden low-frequency sounds by resonating. A material's ability to reflect or absorb sound is described as its absorption coefficient, where a perfect absorber would have a coefficient of 1 and a perfect reflector a coefficient of 0. The construction industry uses a system of measurement called the noise reduction coefficient (NRC), which is an aggregated measurement of absorption across a range of frequencies.

Sound Insulation

Mass is critical for insulation. Sound insulation is often heavy and dense as opposed to thermal insulation which may be lightweight and air-filled. By using heavy, massive exterior and party walls you can prevent unwanted sound transmission. The sound insulation of a single panel is proportional to its mass per unit area. This is known as the mass law. By doubling the mass of that panel you will increase the sound reduction index (SRI) by 6 dB. Most construction materials have an estimated SRI, which in the design of a new building or space can be aggregated together to give you a composite sound reduction index for that space. New construction methods and specification techniques can, however, create lightweight solutions with good acoustic insulating properties for walls and partitions. Dual panel partitions separated by a hollow air cavity will improve on the identical mass of a single panel; this is called an "increase over mass law." If the two panels are structurally isolated with separate framing we can further improve acoustic insulation. By further increasing the air gap and adding acoustically absorbent material such as glass fiber the performance of this lightweight construction can begin to compete with that of a massive monolithic construction. Sound insulation and control is also described as sound attenuation.

Acoustic Isolation

Another way to prevent sound transmission from one space to another is to isolate a structure or room acoustically. This is achieved by creating a floating room within a room with the space separated from the main structure by a series of acoustic isolators (shock absorbers). This acoustic separation is continued around structural openings such as doors or windows by using rubber or neoprene gaskets. In the same way that you may introduce a thermal break to prevent "cold bridging," this acoustic break means that the sound from one space will not be able to travel to another. This technique is often used in the design of sound recording studios and machine rooms to prevent vibration (low-frequency sounds) affecting other inhabited rooms.

1　Reflection of sound.
2　Parabolic reflector "focusing" sound.
3　Whispering Gallery, St Paul's Cathedral, London, UK.
4　Table showing noise reduction coefficients (NRCs) for building materials.

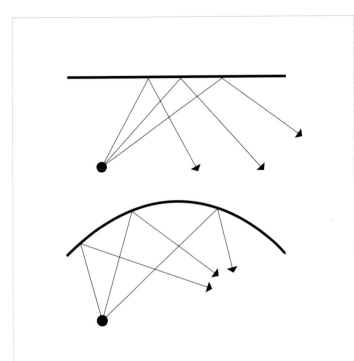

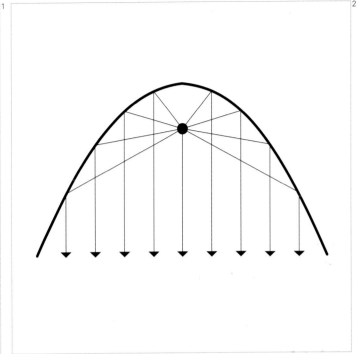

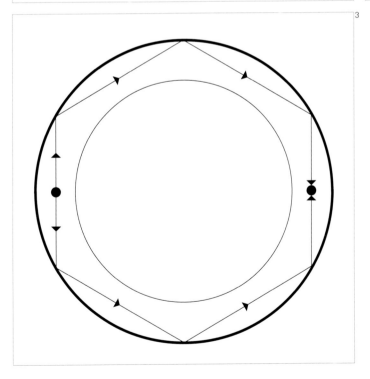

Noise Reduction Coefficient	(NRC)[4]
Brick	.00–.05
Carpet (with underlay)	.30–.55
Concrete (smooth)	.00–.20
Glass	.05–.10
Plaster	.05
Plywood	.10–.15
Rubber on concrete	.05
Seating occupied	.80–.85
Seating unoccupied	.30
Steel	.00–.10
Terrazzo	.00
Wood	.05–.15

106 Sound and Light / Daylight

When designing a building, the maximum use of available daylight can contribute to both the well-being of its occupants and to energy conservation.

Biological Effect of Daylight

Light sends signals via the novel photoreceptor to the biological clock that regulates our circadian rhythms. Light can trigger release of the stress hormone cortisol and the sleep hormone melatonin. Without enough natural light, humans can have problems functioning.

Daylight Factor

There are two separate components to natural day lighting: daylight – diffuse light from the sky; sunlight – direct-source light from the sun.

The average daylight factor for a room can be calculated according to the room's location and its dimensions relative to the window opening. The overall daylight factor for a room is made up of three components: the sky component, the externally reflected component and the internally reflected component. To maximize internally reflected light, surfaces should be as light as possible – at least minimum reflectances of 70 per cent for the ceiling, 50 per cent for the walls, and 30 per cent for the floor. A room will be lighter at the back if the following conditions are met: its depth is not much greater than its width; its depth is not too many times the height of the window head above the floor; the surfaces at the back of the room are light.

Glare

Direct glare stems from light sources that are directly visible in the visual field. It can result from daylight, sunlight, artificial light or specular surfaces. It is known that glare limitation prevents errors, fatigue, accidents and discomfort arising from high contrast levels – the eye can adapt to lower levels of light so long as these are uniformly spread throughout the space.

Modeling Daylight

The combined facilities of an artificial sky and a heliodon can act together as a physical simulator to model daylight and sunlight penetration into a building or across a site. An artificial sky comprises a hemispherical dome that supports an array of individually controlled fluorescent luminaries. It can model all types of sky conditions around the world. A heliodon can model the sun's path in relation to a site for any time or location around the world.

1 Calculating the overall daylight factor for a room.
2 Beams of light are created by tiny dust particles.
3 By placing a model building on a heliodon's flat surface and making adjustments to the light/surface angle, it is possible to see how sunlight would penetrate the building and cast shadows at various dates and times of day.
4, 5 Example of an artificial sky.

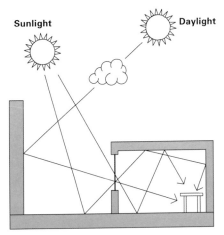

Sunlight　　Daylight

2

4

3

5

Direct Heating

Direct heating systems involve the use of stand-alone appliances in each room of a building. From open fireplaces or wood-burning stoves to electric fires and gas convector heaters,* such appliances require their own direct fuel supply or energy source.

Indirect Heating

Known as central heating, indirect systems rely on a single source – a furnace, solar system, or boiler – to generate heat. This heat is then distributed throughout the building by a heat transfer mechanism – pipes for water or steam (wet indirect heating), ducts for air (indirect warm air heating).

Hydronic Heating

In a conventional radiator system, boilers burn fuel such as gas or oil in order to heat the water contained within them to region of 176°F. The water is then pumped around the building. Heat emitters – surfaces – are used to transfer heat from the supply pipes into rooms by convection and radiation. The surface (radiator) is sized according to the volume of the room, and its temperature can be controlled by a "thermostatic" valve that automatically controls the flow of water according to the room temperature.

Convection can be enhanced by encasing the radiator and allowing the air inside to heat up: cool air enters via a lower grille and warm air exits at the top. A fan can be added to drive the circulation of the air.

Underfloor heating

Floors and walls themselves can be turned into heat emitters by embedding loops of (polyethylene) water pipes. These can operate at substantially lower flow temperatures than conventional radiators due to the large surface areas and, thus, are ideal for the application of solar thermal water heating. Underfloor heating can also be provided by electrical heating elements but this is a very energy inefficient solution.

Indirect Warm Air Heating

A centrifugal fan is used to pass air over a heated surface (usually coils of hot water pipes) and then through ducts into rooms. The air may pass through a filter and may be recirculated through a separate duct using a smaller fan. These systems warm rooms up more quickly than wet systems and may also be used for ventilation purposes.

Indirect warm air heating is the standard in the United States. These systems have the advantage over radiant systems in that they can change the air temperature in a space very quickly and may also provide fresh-air ventilation. Occupants, however, generally do not find the space conditions created by warm-air heating as comfortable as those conditions created by radiant systems.

*Gases that are the product of the combustible fuels used in boilers and gas convector heaters are known as flue gases and require a separate outlet. When gas is burnt in air it produces, among other things, carbon monoxide which is an asphyxiant and must be safely removed from the building through a vertical duct known as a flue. Areas where boilers are kept must also be well ventilated.

1 Hydronic heating system.

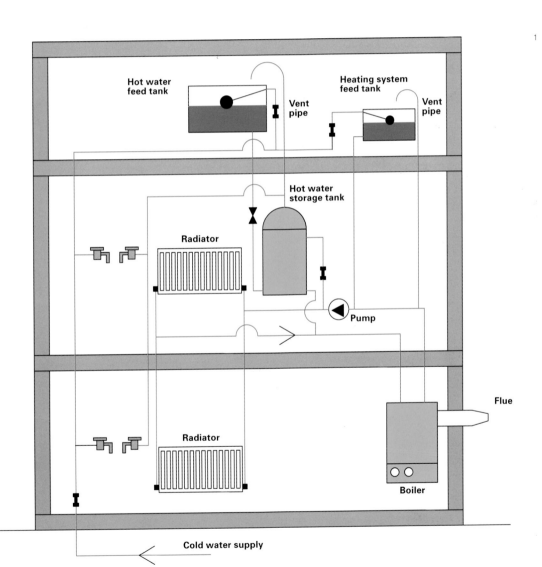

Hot water
feed tank

Vent
pipe

Heating system
feed tank

Vent
pipe

Hot water
storage tank

Radiator

Pump

Radiator

Flue

Boiler

Cold water supply

110 Building Services / Mechanical Ventilation

Indoor air quality (IAQ) has become a major source of concern in the design of buildings. Sources of indoor pollutants include carbon dioxide from building inhabitants, volatile organic compounds off-gassing from building materials, odors from bathrooms and cooking, and moisture from indoor sources. In tightly built modern buildings, these undesirable elements must be removed and fresh air provided to ensure the health, comfort, and productivity of building occupants.

Extract Ventilation

In both domestic and commercial buildings, rooms that generate high levels of moisture or odour, such as kitchens and bathrooms, should have a means of extracting stale air. In bathrooms, this is generally accomplished with an exhaust fan and in kitchens and laboratories with a ventilation hood that ducts the air to the outside.

Supply Ventilation

Buildings should ideally have a regular supply of fresh air provided to the occupants to ensure adequate IAQ. The simplest means of supplying fresh air is via supply ventilation or make-up air. Instead of sucking stale air out of a building, fans are used to force air into it. The effect is to pressurize the internal spaces and drive the stale air out.

Balanced Ventilation

As buildings have become tighter, and energy use a greater concern, the use of balanced ventilation has emerged as the preferred strategy for providing fresh air and removing indoor pollutants. In a similar process to indirect warm air heating, combining supply and extract ventilation in a single circuit allows the system to provide a constant supply of fresh air to a space. If the supply air is cooler than the air inside the building, heat can be recovered from the warm, exhausted air by passing it over a heat exchange element.

Humidity Control

Certain levels of humidity have to be maintained in a building, because of their impact on its fabric and the different types of material being stored within, for the comfort of its human occupants and to control molds and allergens that become active at high humidity levels. An average house produces some 26 pints of water vapor per day and when warm moist air comes into contact with a cooler surface it condenses. This condensation then causes materials to decay, e.g. timber to rot, steel to rust. Encouraging natural air movement through buildings will help to reduce the potential for condensation. Mechanically, humidity levels can be raised using humidifiers or lowered using dehumidifiers. As with heating and cooling, these mechanisms may be applied directly within rooms or within the air handling unit of an air-conditioning system.

1 Ventilation system.

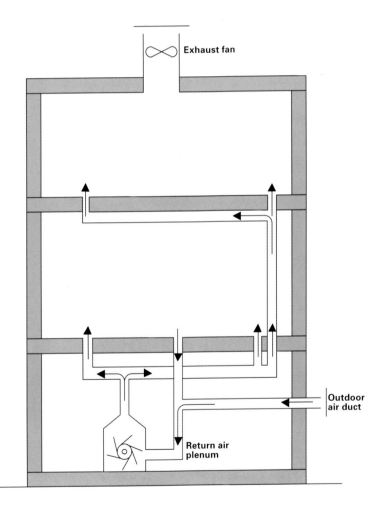

112 Building Services / Air-Conditioning: Principles

Air-conditioning describes a number of systems for modifying the climate within buildings. Usually, it is concerned with cooling, but centralized systems can also provide heating, air filtration,* humidity control and ventilation.

Vapor Compression Cycle
Electrically powered chiller units work on the same principle as a domestic fridge. For a liquid to evaporate it must absorb energy. By passing a refrigerant (a liquid that evaporates easily at relatively low temperatures) through a coil, it removes energy from the coil by taking heat from the air around it: thus the air is cooled. The vaporized refrigerant is then condensed using a compressor and passed through a second coil where the (waste) heat energy released by the condensing process is again passed to the air.**

Absorption Chilling
Absorption chillers use heat instead of mechanical energy to provide cooling. In absorption chilling systems, the mechanical vapor compressor is replaced by a thermal compressor. The process is based on evaporation, carrying heat in the form of hot molecules from one material to another that absorbs them (similar to human sweating where water from sweat evaporates and is "absorbed" into cool dry air, carrying away heat in water molecules). Absorption chilling works in practice by evaporating liquid ammonia and dissolving the vapor in water. This solution is pumped to a generator where the refrigerant is revaporized; the solution is then returned to the absorber.

* Air carries all sorts of suspended particles, from dust and fibers to microscopic bacteria and viruses. With centralized air-conditioning, these particles must be filtered on entering and leaving buildings and periodically cleaned from ducts and from the air handling units. Filters are made from a variety of fabrics of differing porosity and may be impregnated with (anti-microbial) biocides, employ electrostatic attraction or use carbonaceous material to filter gaseous pollutants.

** A heat pump employs a vapor compression system for heating rather than chilling. The cool air is delivered to the outside and the warm air is used for space heating. Reverse cycle heat pumps can provide both heating and cooling as and when required; they can be used very efficiently to transfer thermal energy from the warmer parts of a building to the cooler parts.

1 Vapor compression cycle.
2 Absorption chilling.

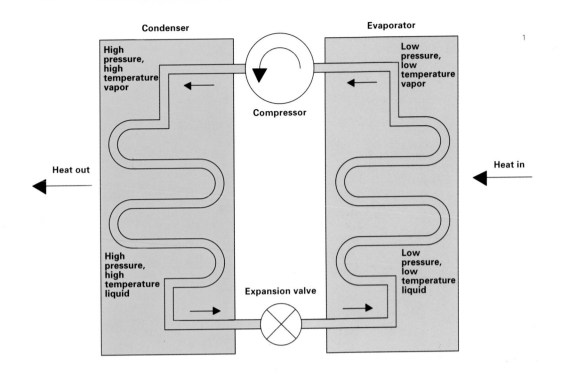

Condenser

High pressure, high temperature vapor

Compressor

Evaporator

Low pressure, low temperature vapor

Heat out

Heat in

High pressure, high temperature liquid

Expansion valve

Low pressure, low temperature liquid

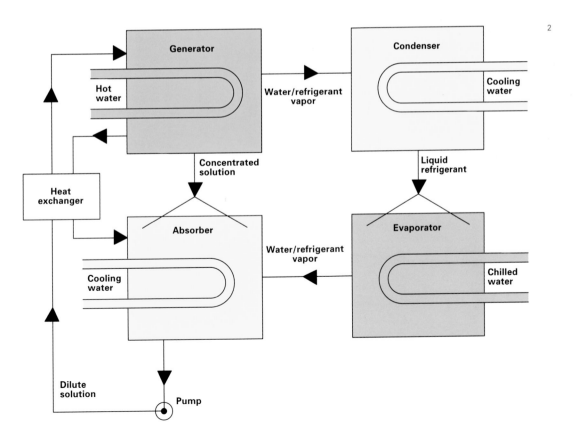

Generator

Hot water

Water/refrigerant vapor

Condenser

Cooling water

Concentrated solution

Liquid refrigerant

Heat exchanger

Absorber

Cooling water

Water/refrigerant vapor

Evaporator

Chilled water

Dilute solution

Pump

¹¹⁴ Building Services / Air-Conditioning: Systems

Evaporation Cooling

An evaporative cooler (often called a swamp cooler) is essentially a large box- like frame containing a big fan that drives the hot outside air through pads that are continually soaked by a water pump. This cools the air by about 20°F as the air evaporates water molecules from the pads. The fan then blows the water-cooled air through the building. In hot dry climates, evaporative cooling is a very energy-efficient and cost-effective solution. It is not suitable for humid climates because of the excessive moisture vapor already in the air.

Unitary Systems

Both the evaporator and condenser are contained within a single unit in a unitary air-conditioning system. This type of system can range from the window unit to a ducted rooftop system. Window or through-the-wall units knowns as packaged terminal air conditioners (PTAC) provide local cooling to the space in which they are located. A fan draws air from the room, through a filter, and over the evaporator coil before discharging it back to the room. Outside air is drawn over the condenser coil to remove waste heat. Local cooling units are generally less efficient than larger systems, but can sometimes result in less total energy use since they may target only a specific space that needs cooling at a given time.

Ducted unitary systems (often referred to simply as package units) work similarly, but with greater capacity and can distribute air through ducts to remote locations. Package units are most often mounted on the roof of a building and discharge air downward into the space.

Split Systems

In a split system, the condenser is located outside and the evaporator inside while refrigerant is piped back and forth between them. This is the most common design for residential air conditioning systems in the Unites States. Inside air is filtered (and mixed with a proportion of tempered fresh outside air in the better systems), drawn over the cooling coil, and returned to the space via ducts. The outside unit discharges the heat absorbed from inside by the refrigerant to a heat sink, most often outdoor air. A central air-conditioning system can provide heating, cooling, air filtration, humidity control, and ventilation through one central air-handling unit that blows conditioned air through ducts to the entire zone that it serves. Air typically returns at a single point to the air handler to be circulated once more. Most residences consist of only one zone, but larger buildings may have many.

Mini-split Systems

In most cases, mini-split systems work much like a conventional split system, except that the inside unit is not connected to ducts. Such a system takes air in and discharges it from the evaporator unit itself relying on velocity and convection to distribute it through the space. An advantage of these systems is that 4 or more evaporators may be connected to a single condenser unit up to 150 ft away, providing an effective solution for zoning the building so that individual spaces may be conditioned only as necessary.

The evaporators for mini-splits may also be concealed in a ceiling or connected to short duct runs.

Geothermal Heat Exchange

If available, other media make much more efficient heat sinks than outdoor air. Bodies of water such as ponds, swimming pools, lakes, and seas provide excellent heat sinks (or sources in the heating season) where heat absorbed by the refrigerant may be discharged by the condenser. The earth is also an effective heat sink/source in climates with relatively balanced heating/cooling seasons, but it is usually quite costly to bury adequate piping to make use of the geothermal stability of the earth's underground temperature.

1 Centralized air-conditioning system.
2 Air-conditioning fans.
3 Active/passive cooling systems. Natural air flow is enhanced by mechanical fans and exhausts. However, the cool temperatures found below ground are employed to cool the incoming air.

1

Outdoor unit

Cool air

Cooling unit

Heat
removed
from
indoor air

Refrigerant piping

Compressor

Warm return air

Filter

Furnace

2

3

Warm

Wind turbine

Fan

Cool

116 Building Services / Integrated Systems

In terms of the environment, the aim of all construction projects is to minimize the amount of carbon emissions both during construction and for the life of the building. This means exploring both the sourcing and transportation of materials and the construction process itself – the embodied carbon in a building – as well as its long-term environmental strategy.

In practice, most buildings employ environmental strategies that combine both passive and active components, viz.:

Natural ventilation

Natural ventilation can be optimized by automatically opening and closing windows, rooflights and/or dampers via an integrated energy management system, however, while opening a window does provide ventilation, the building's heat and humidity will then be lost in the winter and gained in the summer.

Mechanical ventilation

Where mechanical ventilation is used to solve this problem, it can be integrated with air conditioning and evaporative cooling systems to cool below ambient temperature* or with heating and heat recovery systems to raise internal temperatures.

Solar shading

Automated louvres can reduce cooling loads in summer by preventing solar heat gains and reduce heating requirements in winter by closing to reduce heat loss at night time.

Photovoltaic cells may be integrated so as to generate electricity.

Ground source heat pump

Since the temperature below the top 20 ft of the earth's surface is nearly constant at between 50° and 60°F, a ground source heat pump uses the earth as a heat source in the winter or a heat sink in the summer by employing a heat exchanger in contact with the ground. The simplest method pumps refrigerant through coils or loops of copper tube that are buried in the ground.

Combined heat and power

Combined heat and power (CHP) systems make use of the exhaust heat (e.g. steam) from electricity generators by capturing it and feeding it back for heating or hot water. Most often used in district heating, where for reasons of efficiency heat is generated in a central location for a number of buildings (rather than each having its own boiler).

*Combined evaporative cooling and ventilation systems only require a small quantity of electricity for the fan that circulates the air. They are particularly effective in large spaces – industrial or commercial buildings, such as warehouses – where a pure ventilation system is unable to provide a suitable internal climate.

1 Heat Recovery Ventilation systems (HRVs) introduce fresh air to a building, and heat it using the warm air that is being removed.

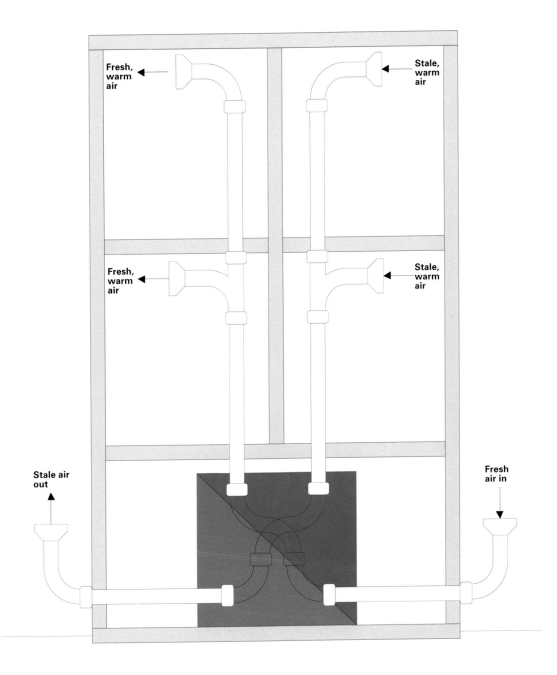

The World Health Organization states that: "Sanitation generally refers to the provision of facilities and services for the safe disposal of human urine and faeces."

In practice, architects are concerned with sanitation in relation to the organization of both the supply of clean water and the disposal of wastewater, as well as the provision of refuse and recycling amenities.

Supply

Water supplied directly to buildings from reservoirs via (normally underground) piping is commonly known as running water or simply tap water. Depending upon the source, this may or may not be drinkable. Water for cleaning (as opposed to drinking) may also be supplied directly from rainwater-collecting cisterns or via on-site holding tanks.

Wastewater collection

In urban areas, the collection of wastewater is normally via sewers for treatment in wastewater treatment plants (for reuse) or for disposal in rivers, lakes, or the sea. Sewers in older, urban areas are often combined with storm drains, and heavy rainfall can lead to raw sewage being discharged into the environment.

In suburban and rural areas, households that are not connected to sewers discharge their wastewater into septic tanks (cisterns) or other types of on-site sanitation. Depending upon space and ground conditions these can either be sealed containers that are periodically emptied or caissons (watertight retaining structures) that permit the effluence to degrade and soak away into the ground below (paper is disposed of separately).

Gray water

Gray water is wastewater that does not contain faecal pathogens. If composting toilets are employed to provide for the separation of urine and faeces at source, all wastewater will consist only of gray water, which can be recycled for gardening or safely discharged into sewers.

Above-ground waste

Traditionally waste pipes were made from metals such as cast iron, lead and copper, though these days pipes are made from a variety of plastics and in standard sizes of 10 mm (3/4 in) to 110 mm (4 1/3 in) diameter according to function. The most widely used are: ABS (Acrylonitrile Butadiene Styrene), MuPVC (Modified Unplasticized Polyvinyl Chloride) and PP (polypropylene).

To prevent sewer gases from entering buildings, a U-shaped pipe (or similar) is always connected between the fitting (bath, sink, toilet, etc.) and the waste pipe. Known as traps, the U-bend always retains a small amount of water and creates a seal that prevents sewer gas from passing from the drain pipes back into the building. Toilet pans have built-in traps. Waste pipes carrying urine and faeces (known as soil pipes) must be independently vented to beyond the roof line (a soil stack).

Drainage systems

Below ground, the waste pipes must connect to drainage systems that in turn are connected to a sewer or other wastewater receptacle. Drain pipes can be made from a variety of materials (most commonly, clay, PVC-U, polypropylene or a type of synthetic rubber known as EPDM) in sizes ranging from 100 mm (4 in) to as much as 600 mm (23 1/2 in) in diameter for major sewers.

A critical part of all wastewater systems is the ability to access, inspect and maintain the pipework both above and below ground, and hence the architect must make provision for access panels, manholes, etc. as appropriate.

1 Typical domestic supply and waste diagram.

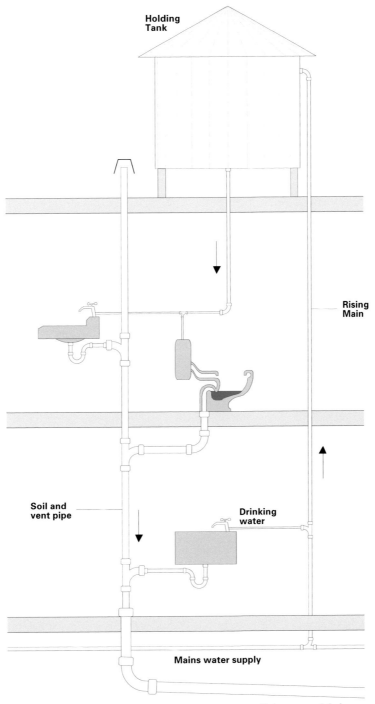

Holding Tank

Rising Main

Soil and vent pipe

Drinking water

Mains water supply

Below-ground drainage

Computational Tools and Techniques

Although some would credit the abacus, the first computational machine was probably a mechanical calculator designed by the French mathematician Blaise Pascal in 1642. In 1804, Joseph Marie Jacquard designed the Jacquard loom, a pattern-making tool that could produce a variety of patterns by simply changing a template. This loom was the first (re)programmable machine to make use of punchcards as a way of encoding data.

Known as a "difference engine" and consisting of sets of gears, levers and pulleys made from cast bronze, the first fully (re)programmable computation machine was designed by Charles Babbage in 1830, but it wasn't until the 1940s that the modern computer was conceived. During the Second World War, Alan Turing, in his search for ways of cracking the German secret code, conceived the idea of an endless tape loop being fed into a machine capable of performing an infinite number of calculations – the "universal computer."

The often used acronym CAD originally stood for computer-aided-drafting. "Drafting" was replaced by "design" at some later stage, and although graphic computing has proven a useful aid to the draftsman, the ability of the computer to help with design is largely a product of its ability to produce rapid mathematical calculations (largely using ones and zeros – the binary system) and hence to (1) organize vast amounts of information and (2) simulate (model) the predicted behavior of both natural and artificial objects and systems.

This section is subdivided into three main categories:

Building Information Modeling
The ways in which computers can be used as interoperable libraries and databases and how building information can be modeled in a computer.

Structural Analysis
The ways in which computers can be used to predict the behavior of materials and structures under load and to explore structural forms within these parameters: form finding and finite element analysis.

Environmental Analysis
The ways in which computers can be used to predict environmental behavior and to explore the design of climatic enclosures within such parameters, viz.: thermal, daylight, air movement, and acoustic.

Above left
Finite element analysis (FEA) of an anticlastic shell structure.

Below left
Form finding – recombining 3D "hyper" shapes.

Building Information Modeling

Building Information Modeling (BIM)

Also known as BIMM – Building Information Modeling and Management – BIM is a generic term that covers a multitude of modern, computer applications. BIM relies entirely on software design and the ability of the different programs to become integrated within a single computer model.

A building information model represents the building as "an integrated database of co-ordinated information." Standard computer aided drafting software is adapted to include object oriented systems. That is to say that since any line on a drawing is able to be encoded as an object, it is then a logical step to link objects in an architectural drawing to non-graphical data using standardized computer file formats. Any construction element in the drawing can also carry other (hidden) attributes such as material properties, costs, suppliers, etc. This non-graphical data, held in a series of spread sheets, can then be extracted for use in written schedules and specifications.

The following range of programs may contribute to the overall model:

Site surveys: 3D geo-technical surveys, thermal imaging, noise mapping, wind mapping.

Design: parametric modeling – an important part of an efficient building information model, parametric modeling is computer software that uses a relational database within a dynamic model of the whole project. The software takes account of the linked behavior of the different elements of a building – its parametric components – both graphically and informationally: it maintains consistent relationships between elements as the model is manipulated. The changes (to a floor plan, a section, a schedule) made in a project are reflected throughout the project and all necessary adjustments and alterations are made automatically.

Structural analysis: materials selection software; finite element analysis (FEA).

Environmental analysis: solar geometry, daylight penetration, thermal capacities, acoustic properties, air movement, and crowd control – the latter two using Computational Fluid Dynamics (CFD).

Optimization and clash detection: the parametric model is also able to identify all building elements on the 3D drawing and use them as a database for schedules, spreadsheets, and specifications. It can therefore help calculate detailed material quantities and track material quantities in cost estimates.

Construction: off-site (pre-) fabrication of construction elements using Computer-Integrated-Manufacturing where machine codes are converted directly from the designer's graphic files; the contractor's program, critical paths, sourcing, and costs etc. are incorporated into the optimization model. (The industry is establishing a standard specification format for BIM.)

Clients: BIM for clients is viewed on line as non-executable files that do not require the original software application; Facilities Management (FM) is able to use the model after hand-over to locate, repair, or replace any element in the building. Bar-coding of elements is also used for this purpose.

1 Centralized computer model.

2 Parametric design using Revit software.

2.1 The 3D nature of BIM means one change is a change everywhere; plans, sections, details, schedules, and quantities will all automatically update.

2.2 "Families" are the 3D components used in the model and can include cabinetry, equipment, building parts, walls, and columns. These components are completely parametric, allowing an increased level of flexibility and reducing the time spent updating traditional plans, sections, and details.

2.3 BIM allows for the project team members including consultants, client, specialists, suppliers, and contractor to work on a single shared 3D model which encourages collaborative working relationships and reduces the risk of construction conflicts and the cost and time impacts of redesign.

2.4 Projects can be visualized at an early stage giving clients a clear idea of the design intent. Before works on site commence BIM allows the project team to work through the "build" in a virtual environment to optimize the construction period.

2.5 BIM enables much faster and more efficient project delivery with the 3D model able to produce quantities, schedules, and fabrication and construction drawings.

2.6 BIM models can contain product information that assists with the ongoing operation and maintenance of a building once completed.

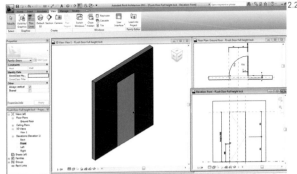

2.1

2.2

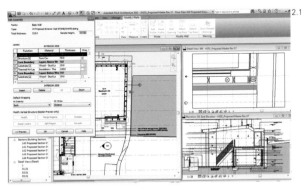

2.3

2.4

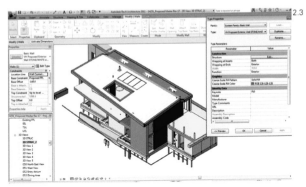

2.5

2.6

Architect

Visualization

Structural engineer

Energy analysis

BIM

MEP engineer

Specifications

Contractor

Owner

Responsive Systems

Computers may be used to enable buildings to adapt and respond to external conditions through electronic sensing and control and the mechanical actuation of their parts; the capacity for the built environment to respond and adapt in a controlled manner depends upon automotive systems.

Automation in buildings requires combined mechanical and electrical systems (M & E) and these range from programming the pumps and fans of heating and ventilation systems through to the control of lifts and the monitoring of fail-safe mechanisms. A modern office building will have a series of computer-controlled routines to monitor and regulate the levels of temperature, humidity and light, as well as air quality and other hazards. Modern, long span/high traffic bridges have built-in vibration and seismic sensors which are monitored 24 hours a day.

Sensors and Actuators

Sensors can monitor the environment in a variety of ways. They can detect motion, temperature, humidity, wind speed and direction, light levels, noise levels, air quality, pressure, vibrations, etc. By turning analogue waves into digital bytes, they can provide a constant stream of data that can either be stored for analytical purposes or fed directly into a program that will search for certain parameters within which to react. These programs may be considered as "if – then" routines: if the sensor detects a certain type of data, then it can instruct a motor to actuate in a certain way. Electronic stepper motors can be computer-controlled to operate in digital "steps" to very fine levels of accuracy.

A to D and Binary Logic

Generally, the type of input that might be required in order to modify an environment may be classified as either a simple switch, such as a light switch, or data that comes in the form of an electromagnetic wavelength such as sound, light or temperature. Sensors that are designed to read this type of environmental data must then transform the analogue waves into a digital format – A to D – so that electronic logic systems can understand and act on the information.

Digitization has developed using binary logic; all data can be encoded using only two symbols (e.g. one and zero) that can then be transformed into a voltage that is in turn directed through logic gates in order to control the desired output.

Expert Systems and Statistical Analysis

In the Case Studies section at the end of this book, two projects illustrate the use of responsive systems in architecture. Foster Associate's HSBC building in Hong Kong (see p168) employs mirrors to track the sun in order to reflect light into the atrium; this is known as an expert system, as the computer knows where the sun is at all times and can adjust the mirrors accordingly. The case study of the "D" Tower by Nox Architecture (see p178) illustrates the use of a regularly updated statistical database to control the external appearance (or "mood") of the building.

1 Binary logic tables are also known as truth tables.

2 Sound wave described using a 4-bit A to D convertor. The analogue wave will be digitally encoded as 1001 1011 1101 1100 1011 1000 0101 0010 0001, etc.

3 Logic gates: graphic symbols and truth tables.

4 Arab Institute, Paris, France. This building by Jean Nouvel employs the same mechanism as that used for camera shutters to control light entering the library. Sensors monitor light levels and automatically open and close the shutters to maintain the optimum levels.

5 Kingsdale School, London, UK (dRMM Architects). The ETFE (ethylene tetrafluoroethylene), inflated "pillows" that cover the open forecourt have an intermediate layer that is printed with the inverse, checkerboard pattern to that of the upper skin. When sensors detect overheating the pressure within the pillows is regulated to bring the two patterns closer together, thus decreasing solar gain.

6 Dynamic yacht mast. Rather than engineering for the worst case scenario, a lighter mast can react to increased loading through sensors and control systems that actuate tension wires within the mast.

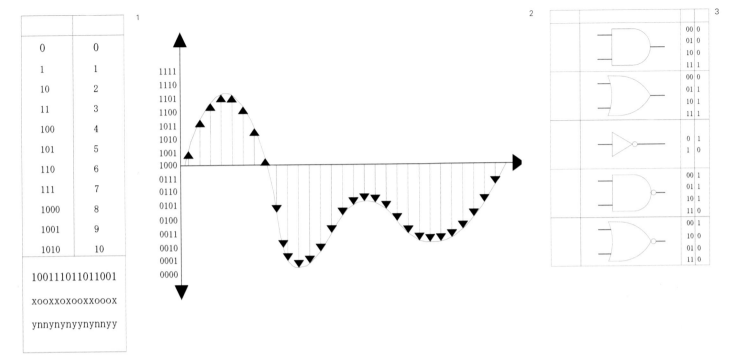

126 Structural Analysis / Form Finding, Finite Element Analysis

Form Finding

Historically, finding and creating new structural forms was accomplished by extracting geometric information from physical models, in particular three-dimensional compressive surfaces (shells) or three-dimensional tensile surfaces (membranes). With the advent of computer-aided-design (CAD) along with an increased knowledge of the behavior of materials, a variety of approaches to form finding can now be pursued using computer programs to calculate optimum structural solutions for given geometric parameters.

Finite Element Analysis

The first step in using finite element analysis (FEA) is constructing a finite element model of the structure, to be analysed. Two- or three-dimensional CAD models are imported into an FEA environment and a "meshing" procedure is used to define and break the model up into a geometric arrangement of small elements and nodes. Nodes represent points at which features such as displacements are calculated. Elements are bounded by sets of nodes and define the localized mass and stiffness properties of the model. Elements are also defined by mesh numbers which allow reference to be made to corresponding deflections or stresses at specific model locations. Knowing the properties of the materials used, the software conducts a series of computational procedures to determine effects such as deformations, strains and stresses that are caused by applied structural loads. The results can then be studied using visualization tools within the FEA environment to view, and identify the implications of, the analysis. Numerical and graphical tools allow the precise location of data such as stresses and deflections to be identified.

1 3D "Nurbs" model of the canopy for a bandstand, with computer-generated section lines highlighted in yellow.

2 The complex roof geometry for a new roof on an existing tower was rationalized using three-dimensional models, and built with a simple steel ring beam and curved steel cross beams that support rafters and a double curved plywood deck.

3–5 Form-finding software used for the design of membrane structures. Control points (CPs) are used to create space and the program operates in such a way that when a force is applied to one point the load of the force is distributed homogeneously so that the membrane is always under tension to produce a smooth transition between points.

6 View of the mesh used to define the geometry of the bandstand canopy.

7 Exploded drawing of the structural components for the canopy.

8 FEA model of buckling in a proposed 230 ft high fiber-reinforced plastic (FRP) mast.

9 In this project for a 32 ft 9 in high tower (designed as a stack of solid acrylic blocks), FEA was applied in order to predict the behavior of the towers under wind load.

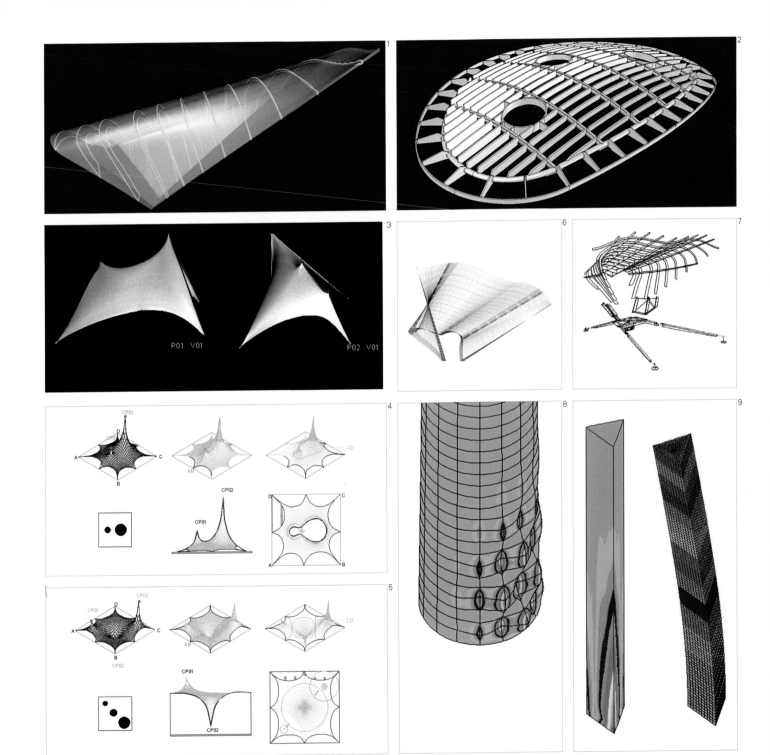

Computer programs can enable architects to carry out environmental analysis for thermal comfort, daylighting, air movement, and acoustics. A three-dimensional graphic model is given material attributes and software produces both numerical and graphical results.

Thermal Comfort – Shading Design and Solar Analysis

Excessive solar exposure is one of the main causes of overheating in buildings, even in relatively cold climates. At the same time, the sun is one of the most effective sources of natural energy available. Thus, shading systems and the analysis of solar gains are inextricably linked.

Analytical computer software gives the architect the ability to calculate and visualize incident solar radiation on the windows and surfaces of buildings over any period of time. The computer model can display overshadowing from adjacent buildings and make it possible to compare incident gains during different seasons, showing variations in the solar resources available at times when heating or cooling is required. This enables architects to quantify the effects of different solar shading approaches, including building orientation and the size and location of window openings, at the design stage.

Sample Thermal Analysis Model

The effect of thermal mass is modeled by placing a building in Athens and applying the inbuilt weather files. The single-story building is 20 by 33 by 8 ft high and the envelope can be constructed from either dense, high thermal capacity materials such as brick and concrete or lightweight, low thermal capacity materials such as timber frames and suspended timber floors. Although the outside temperature reaches 103°F at midday, the heavyweight building maintains a 9.7°F temperature difference at this hottest time of the day. The maximum internal temperature of 94.8°F occurs at 4 pm (diagram **1**). The lightweight building shows the effects of instantaneous heat gains with a maximum temperature occurring nearer to midday, at 2 pm, with a higher internal temperature of 97.3°F and a maximum temperature difference of only 6.8°F. When comparing heavyweight and lightweight buildings, the concrete and brick envelope reduces the August cooling power required to 45 per cent of that of the timber-frame structure. If cooling were to be achieved by air-conditioning with refrigeration, the August running costs and carbon emissions for the concrete structure would be 70 per cent of those for the timber frame. By making the Z axis value on the original drawing −8 ft, the effect is to bury the entire zone in the earth. The resulting monthly load graph clearly shows that the maximum cooling requirement has reduced to 592W (diagram **2**): there is therefore no requirement to heat the building.

Daylighting

Because analytical computer software can automatically generate sun path diagrams to show overshadowing periods for the entire year, it is possible to calculate daylight factors and illuminance levels at any point within a model (see diagrams **3–6**).

1, 2 Thermal analysis bar charts.
3 Daylighting analysis; location London, shadows at 4 pm.
4 Daylighting analysis; view from sun's position.
5, 6 Daylighting analysis; daylight values and penetration into interior.

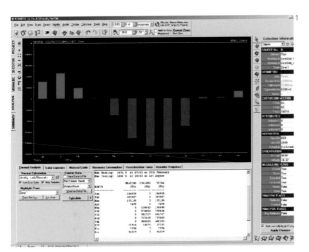

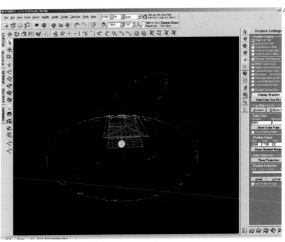

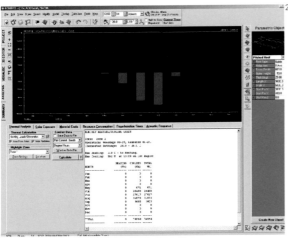

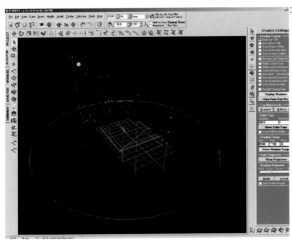

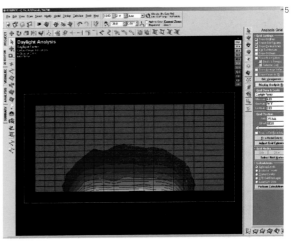

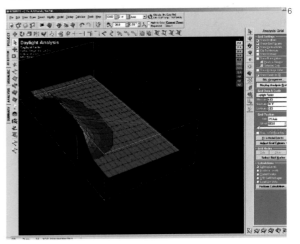

Environmental Analysis / Air Movement, Acoustic

Air Movement

Using inbuilt weather data, analytical computer software can overlay annual wind speed, frequency and direction directly on top of a design model, making it appropriate for natural ventilation and wind shelter strategies, particularly when combined with computational fluid dynamics software.

Computational Fluid Dynamics (CFD)

The Navier-Stokes equations, named after Claude-Louis Navier and George Gabriel Stokes, are a set of equations that describe the motion of fluid substances such as liquids and gases. The equations are a dynamical statement of the balance of forces acting at any given region of the fluid. The various numerical approaches to solving the Navier-Stokes equations are collectively called computational fluid dynamics. When translated into a graphical format, the motion of the fluids can be seen as particles moving through space. CFD can then be used to simulate wind dynamics – speed and direction – in and around buildings. The architect can explore variations in design that can, for example, improve natural ventilation or minimize excessive down-drafts from tall buildings.

Acoustic Modeling

Analytical computer software offers a number of acoustic analysis options. These range from simple statistical reverberation times through to sophisticated particle analysis and ray tracing techniques.

Sample Acoustic Analysis Model

The effect of different materials and surfaces on reverberation time is explored within a computer model. The single-story building is 20 by 33 by 8 ft high and two different types of envelope are analyzed. In the first, the envelope is constructed from masonry cavity walls with a plaster joist ceiling and concrete slab floor. This gives reverberation time data of 2.3 seconds at 500 Hz and 1.14 seconds at 1 kHz (diagram **2**). In the second, the envelope is constructed from masonry cavity walls with an acoustic tile ceiling and a carpeted suspended timber floor. This gives reverberation time data of 0.9 seconds at 500 Hz and between 0.31 and 0.75 seconds at 1 kHz (diagram **3**). The graphical outputs clearly show the impact of absorbent surfaces on sound reflections within the room.

1 CFD applied to assess wind impact on a façade.

2, 3 Acoustic reverberation graphs.

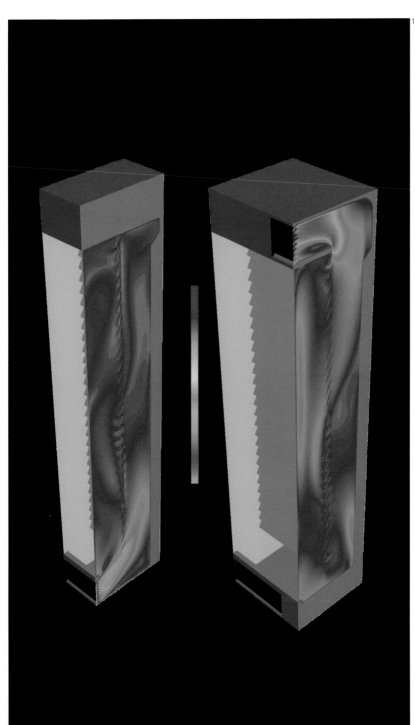

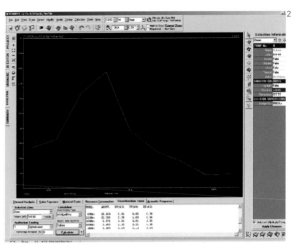

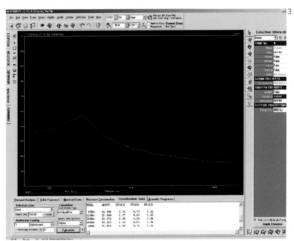

Case Studies

Origins of Construction Types

The following range of case studies illustrates the historic relationship between design and technology in architecture and civil engineering. The examples are listed chronologically. Generally, the older the structure, the more likely it is to be low tensile in nature – the materials used are largely in compression and are therefore unsuitable for long spans. At the same time, older structures were not necessarily less environmentally efficient than modern ones, and it can be seen that passive environmental control was inherent in much traditional, vernacular architecture. Traditional dwelling types were born out of the local environment – both making use of its physical resources and reacting to its climate. Whether it was trees, rocks, mud or snow, each material has found a way to be employed as both structural form and climatic envelope. The first reliable traces of human dwellings, from as early as 30,000 years ago often consisted of a circular or oval ring of stones, with evidence of local materials being used for a tent-like roof. In wet areas, such materials may have been reeds daubed with mud or, in the open plains, mammoth bones and tusks lashed together to support a covering of hides.

Size

Although the circular plan worked as an efficient generator of structural form for many materials and construction methods, tall and long-span structures really began to develop with advances in materials science and industrial processes that were often pioneered in the transport industries. Knowledge from the shipbuilding, automobile and aerospace industries, where the need for strength and lightness has been paramount, was adapted and transferred for use in construction technology. Chain and wire rope, for example, not only enabled the construction of suspension bridges, but also that of lifts without which there would be no skyscrapers.

Prefabrication

Off-site manufacturing, the ability to fabricate building elements in the controlled environment of a factory or workshop, has long been known to result in better quality control. With the advent of the Industrial Revolution, the capacity to transport ever larger components has progressively expanded the possibilities for prefabricated building elements. The ultimate expression of this is the international space station whose parts are delivered in the hold of a space shuttle.

Note: Icons in the case studies cross refer to the main technical elements used in the construction of a structure or building, and also indicate where thermal comfort, air movement, and acoustics have been major factors in its design.

Left
Montage of the case studies.

134 Mud House

> 36 Masonry

> 60 Ribbed Dome

> 72 Thermal Comfort

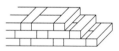

Location: Middle East
Date: From 8000 BC
Height/Diameter: Varied
Materials: Mud bricks

Evidence of bricks shaped from mud and baked hard in the sun can be found as long as 10,000 years ago. The houses were circular in plan and built with bricks that were curved on their outer edges. The floor of each house was excavated some way down into the ground, and then both the floor and the brick walls were plastered in mud. The roofs consisted of a conical structure of branches and mud (wattle and daub). Later, around 6500 BC, there is evidence of roofs in the form of domes that had been built up by stepping ("corbelling") the brick courses towards the center – a technique that had its ultimate expression in the Pantheon in Rome some 6,600 years later.

Mud Bricks

A mud brick is an unfired brick made of clay. Usually found in hot climates where there was a lack of timber to fuel a kiln, they were dried in the sun and had a lifespan of no more than 30 years. The Great Mosque of Djenné, in central Mali, is the largest existing structure made from mud bricks.

Adobe

This is a type of mud brick still used today. It is generally used as a bearing system and provides good thermal storage and insulative qualities. Clay and straw are mixed together and pressed into mold to make adobe.

Rammed Earth

Rammed earth construction is a process of compressing a damp mixture of earth (consisting of sand, gravel, and clay, sometimes with an added stabilizer) into a mold (known as the formwork) in a similar process to that of casting reinforced-concrete walls. Traditional stabilizers included lime or animal blood, but modern rammed earth walls usually employ a cement additive. After compressing the earth, the formwork can be immediately removed but the earth will require a period of warm, dry days to dry and harden. Rammed earth walls can take up to two years to cure completely, but once the process is complete they can be readily nailed or screwed into, and they are good thermal and acoustic insulators.

1 A mud house fabricated from dried earth bricks. Wadi Azzan, Shabwah, Yemen.

136 Tepee

> **66** Mast-Supported Membrane   > **58** A-Frame   > **72** Thermal Comfort   > **84** Air Movement

Location: North America
Date: From 4000 BC
Height/Diameter: Varied
Materials: Timber sapling, animal skin
Engineers: Native American Indians

A tepee is a conical tent originally made of animal skins and popularized by the American Indians of the Great Plains from around AD 1700. The skins were both good insulators and remained dry during heavy rain. Importantly for these nomadic tribes, the tepee was portable.

Tepees were traditionally built using between 10 and 20 sapling poles, with an animal skin cover and sometimes an inner lining for ventilation and/or extra insulation. Initially, three poles would be lashed together at one end and erected to form a tripod-like frame. Other poles were then to form a circular shape at ground level and all the poles were then bound together. Once the skins had been attached to the poles, the non-tripod ones were pulled out at ground level in order to complete the tensioning of the whole structure and the skins were pegged to the ground. This anchoring system, combined with the conical shape of the tepee, made it highly resistant to wind load.

Tepees are different from other tents in that there is an opening at the top with flaps that can be adjusted with long poles. When open, the flaps are set at angles to prevent the prevailing wind entering the tent while at the same time enabling the tepee to act like a chimney and so allow for an open fire to be lit within. An inner lining strung from the poles about 5 ft above the ground could be used either to shut off drafts or to direct the flow of incoming air so as to feed the fire or supply fresh air to the tepee's occupants. In hot weather, the lining would be removed and the outer skin would be unpegged and rolled up to encourage cross-ventilation.

1 Typical native American Indian tepees.

138 Igloo

> **36** Masonry   > **60** Ribbed Dome   > **72** Thermal Comfort

Location: Central Arctic, Greenland
Date: From AD 1000*
Height/Diameter: Varied
Materials: Snow
Engineers: Inuit

An igloo, also spelled iglu, is constructed from blocks of snow in the form of a dome. The dome can be raised by stacking blocks of compacted snow without any supporting structure: blocks are cut to shape with a snow saw or large knife and stacked in a helical fashion. Each block is a rectangle measuring about 24 by 48 by 8 in; the best snow to use is from a deep snowdrift of fine-grained snow that has been blown by wind, which compacts the ice crystals. Joints and crevices are filled with loose snow and a clear piece of ice, or seal intestine, is inserted for a window. If correctly built, an igloo will support the weight of a person standing on the roof.

Like animal skins, snow has a low density and is thus a good insulator. While temperatures on the outside may be as low as –49°F, the inside temperature can range from 19°F to as much as 61°F when warmed by body heat alone. In the traditional Inuit igloo, heat from a stone lamp would cause the interior to melt slightly and this melting and refreezing built up an ice sheet and contributed to the strength of the structure.

Igloos are entered via a narrow, semi-cylindrical passageway about 10 ft long that is sunk into the snow and which then penetrates the igloo through a hole beneath the wall. This allows cold air to exit as the igloo warms up. A raised sleeping platform is constructed in the warmer, upper part of the dome. The Central Inuit line the living area with animal skins, which can increase the temperature within from around 35°F up to 50–70°F by increasing heat retention.

It is vital to make at least one air hole in the roof of an igloo to avoid suffocation, since without ventilation lethal carbon dioxide will build up. An experienced Inuit can build a snow igloo in between one and two hours.

*It is not known when or where the igloo originated, but it is known that the Inuit first migrated to what is now northern Canada roughly 1,000 years ago. These Inuit were descended from Palaeo-Eskimo people who first appeared in western Alaska between 5,000 and 4,000 years ago.

1 Constructing an igloo. After the first row of blocks has been laid out in a circle on a flat stretch of snow, the top surfaces of the blocks are shaved off in a sloping angle to form the first ring of a spiraling helix. It is believed that the helical shape aids in the construction process as successive blocks can lean on each other.

2 A computational fluid dynamic (CFD) model of an inuit igloo, showing the effects of ventilation flows and retained heat.

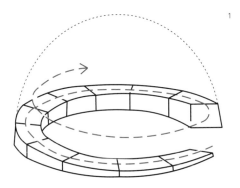

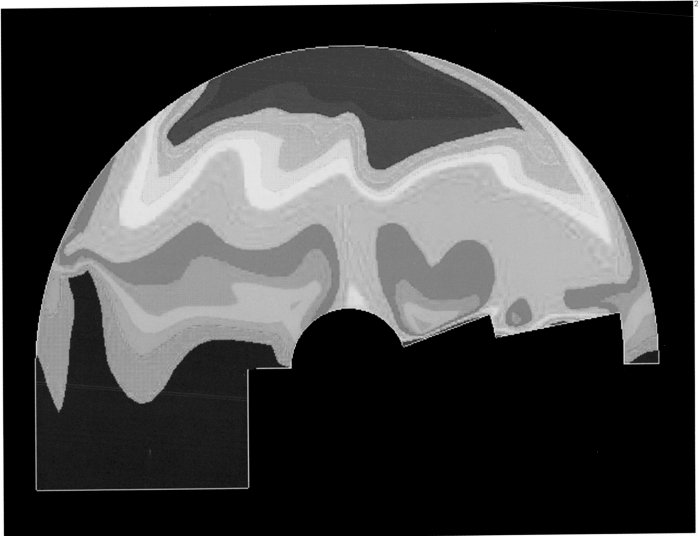

140 Obelisk

Location: Karnak, Egypt
Completion date: 1450 BC
Height/Weight: 97 ft/350+ tons
Materials: Red granite
Engineer(s): Ancient Egyptians

Standing stones (or megaliths) were the product of mankind's first attempts to build tall, slender structures. They are often presumed to have been used as clocks (to cast shadows as with a sundial), and very little is known about the techniques used to erect them. However, there are records that show how the ancient Egyptians made their obelisks.

Often referred to as needles, obelisks were designed to stand at either side (east and west) of the entrance to a temple, to represent the rising and setting of the sun. Much of the granite used for these megaliths was quarried from the hills around Aswan in southern Egypt.

It is not known how the ancient Egyptians could predict that there would be perfect unbroken slabs of granite at the precise locations they chose. However, it is thought that they would have dispatched holy men (seers) to establish where to quarry. Indeed, there is an unfinished obelisk in the hills just outside Aswan that must have developed a crack during the digging of the trenches. It is the strongest piece of evidence for the excavation techniques used and would have been the largest and heaviest of all the obelisks by some margin.

Once a site had been chosen, fires were lit to burn off the foliage and loosen the topsoil.

The rough shape of an obelisk was then marked out and labourers, working in pairs, commenced digging two parallel trenches on each side of it, using dolomite balls attached to sticks in a hammering action. Hundreds of labourers would chant (for rhythm) as they worked, and once the trenches were of sufficient depth, they would tunnel underneath the granite, propping it with lumber as they progressed. After many months, the obelisk would be freed from the ground. It would then be levered out using rough timber logs and transported from the hills above Aswan down to the river Nile where it was placed on a barge. From here it may have travelled as far as Cairo, 400 miles downstream.

Once on-site, the finishing work was carried out. The obelisk carved into a tapering shape in order to lower its center of gravity, the surface of the granite was smoothed, polished and carved with intricate hieroglyphs, and finally the megalith was capped with gold. It is still not known how an obelisks was erected. However, it was placed on a plinth that was dug into the ground to act as a foundation stone.

1 View of the obelisk in situ at Karnak.
2 Detail.

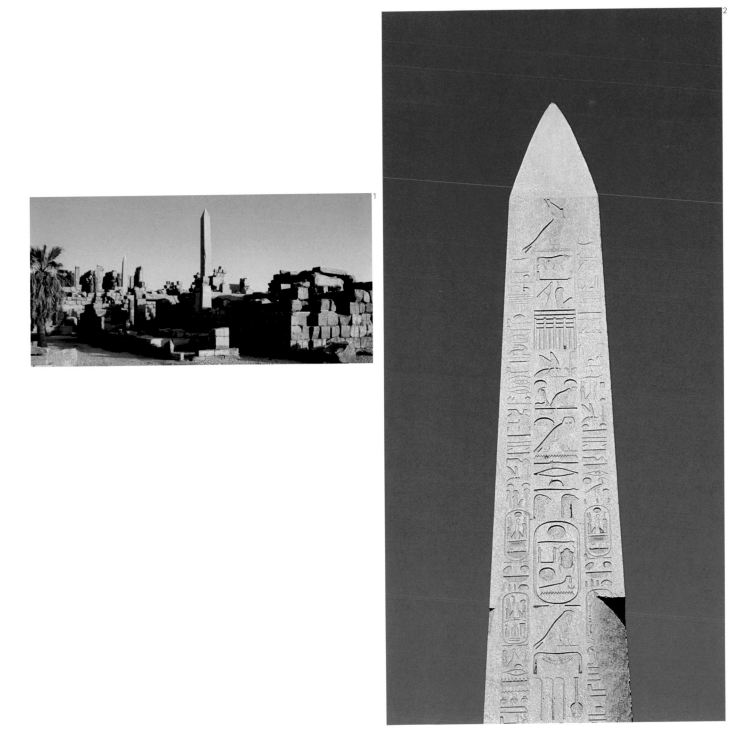

142 The Pantheon

> 36 Masonry

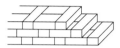

> 60 Ribbed Dome

> 84 Air Movement

Location: Rome, Italy
Completion date: AD 126
Span/Weight: 142 ft/5,000 tons

Built by the emperor Hadrian, the Pantheon was the largest dome in the world until Brunelleschi's dome in Florence in 1436. The dome was built using stepped (coffered) rings of concrete that diminished in both density and thickness as it rose from its base, where it is 21 ft thick, to the oculus where it is approximately 41 ft thick. The dome sits on a cylinder of 21 ft thick masonry walls, which are constructed with a set of eight barrel-vaulted voids and recesses lined with travertine. Although the dome was built with cast concrete, it was not in any way reinforced and still holds the record for the largest unreinforced concrete dome in the history of architecture.

The precise composition of the concrete is not known. Typically, Roman concrete was made from a mix of hydrate of lime, ash, pumice, and pieces of solid rock. However, as there were no reinforcing bars to resist the tensile stresses, it can only be assumed that their concrete was both lighter and better compacted than its modern-day equivalent.

The dome of the Pantheon has a 27 ft circular opening to the sky at its center, which illuminates the vast interior. This oculus also serves as a cooling and ventilation method. As wind passes over the dome it is accelerated, and creates a negative pressure zone that sucks air out from the top of the dome and forces fresh air to be drawn in from the entrance to the Pantheon. The oculus was edged with a circular bronze cornice, which is still in place. The exterior of the dome was originally covered in gilded bronze plates, later replaced by lead.

The internal height of the Pantheon precisely matched that of its diameter in plan, so an imaginary sphere can be envisaged alluding to the cosmos, with bronze stars originally decorating the coffered ceiling and the oculus representing the sun at the center of the universe.

The ceiling of the dome is lined with five horizontal bands of coffered panels that decrease in size toward the top. The moldings of each band are foreshortened to control their appearance from below. The coffered panels reduce the overall deadweight of the ceiling as well as providing decoration. They are lined with marble and porphyry stone.

The portico of the Pantheon measures 110 ft wide by 59 ft deep and features 16 monolithic (one-piece) granite columns imported from Egypt. Each column is 46 ft high with a diameter at its base of 5 ft; this reduces to a diameter of 4 ft 3 ½ in at the column top. Each column weighs an estimated 60 tons. The huge bronze entrance doors and fanlight were originally plated with gold.

1 Plan and cross-section of the Pantheon. The building's interior dome and supporting drum contain a perfect sphere.
2 View of the interior.

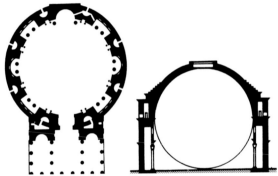

144 The Ironbridge

> **44** Extruded Steel Beam

> **46** Open Web Truss

> **60** Barrel Vault

**Location: Severn Gorge,
Coalbrookdale, UK
Completion date: 1781
Length: 100 ft
Materials: Cast iron
Engineers: Abraham Darby III,
Thomas Farnolls Pritchard**

The Ironbridge at Coalbrookdale in Shropshire, England, was the first bridge (or indeed civil engineering structure) in the world to be made entirely of cast iron. East Shropshire was an important industrial area thanks to coal deposits near the surface, and it was here that the Industrial Revolution was born.

The project was initiated in 1775, when a group including Abraham Darby III, Thomas Farnolls Pritchard and John Wilkinson became interested in creating a physical link between Coalbrookdale and the mines, foundries and quarries south of the Severn River. Thomas Pritchard (an architect) was awarded the job of designing the bridge and in 1777 it was agreed that Darby should build the new bridge with a span of 90 ft. However, Pritchard died later that same year leaving Darby the job of building this extraordinary structure. In 1778 the bridge abutments were complete and in 1779 the giant iron castings were fabricated. The first two ribs were installed in July with each weighing 5.5 tons, and by the fall the main iron superstructure was complete.

The final completion was considerably later, in 1781, after the approach roads were finished. The bridge had a far-reaching impact: on local society and the economy, on bridge design and on the use of cast-iron in building.

Its arch finally spanned 100 ft, and has five arching ribs, each cast in two halves. Although the bridge is made from iron, the construction details were adapted from traditional timber construction, which Darby's workers would have been used to employing. Blind dovetailed joints, where only half the thickness of the iron is in the shape of a dovetail, join the arched ribs to the radials, while mortise joints secured by wedges secure the ribs to the horizontal and vertical members at each end of the bridge.

All the major components were put together in three months without a single accident or the least obstruction to the boats on the river. It is likely that a timber scaffolding would have been employed as a temporary support during the erection of the ribs and to create a temporary span for river traffic to pass beneath.

In the great flood of the Severn in 1795 the Ironbridge was the only bridge left undamaged, which only added to its status as a marvel of contemporary engineering and an advertisement for the possibilities of metal structures.

The completion of the bridge marked out Ironbridge Gorge as one of the most technologically advanced areas in the world at the close of the eighteenth century. This achievement was given worldwide recognition when, in 1986, the Ironbridge Gorge became the first of seven United Kingdom sites to be awarded World Heritage status by UNESCO, to celebrate the area's contribution to industrialized society.

1 Diagram of the blind dovetail joinery that connects the arched ribs to the radials.
2 View of the Ironbridge.

The Bell Rock Lighthouse

> **36** Masonry

Location: Scotland, UK
Completion date: 1811
Height/Weight: 115 ft/2,296 tons
Materials: White sandstone and
Edinburgh granite
Engineers: Robert Stevenson with
John Rennie

The lighthouse was built at a time when increasing international trade meant that many sailors' lives were being lost around the treacherous coasts of Europe. It was built 11 miles out to sea on the east coast of Scotland in the northern reaches of the great sea estuary known as the Firth of Forth, on a reef which, except at low tide, lies submerged just beneath the waves. It was constructed with no power except for men and ponies.

The design of the lighthouse was the culmination of knowledge gained from the construction of previous lighthouses (many of which failed) and from prototyping with scale models. John Smeaton had built the Eddystone lighthouse in 1759, pioneering the use of stone. Not only were the stones dovetailed to interlock with one another but were reinforced vertically with posts (similar to the dowels in a scarf joint). The ideal profile to resist the enormous impact from wind and waves was found to be parabolic in shape. The Bell Rock lighthouse had a broader base than some of the earlier designs, the theory being that, to minimize the action of the sea against a solid tower in such an exposed situation, the sea's force would be better deflected with a shallower rise from the base than if a tower took off at a steeper angle. The first course has a diameter of 42 ft and each of its stone blocks weighs more than a ton.

One of the first things that had to be done was to prepare a vessel to act as a "floating light" – a novel idea at the time –

that would be moored just off the rock. Secondly, a beacon house (a temporary structure that would eventually house the builders) was to be built on the rock. The "Smeaton," named in honor of John Smeaton, was built to act as a tender for the floating light and to transport the blocks of stones from the harbor out to the Rock. Stones had to be transferred from the "Smeaton" to smaller boats and then by special winching tackle onto the rock.

The Bell Rock was submerged twice daily to a depth of sometimes 16 ft, and on average it was only possible to work it about two hours every low tide. In all, 2,835 stones were used in the construction of the lighthouse and the stonework has never needed any kind of repair.

1 Plan of the lighthouse's interlocking foundation stones.
2 Cutaway illustration of the Bell Rock lighthouse under construction.

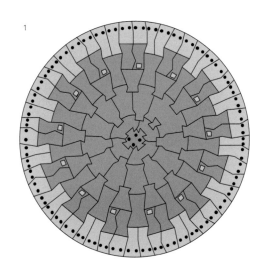

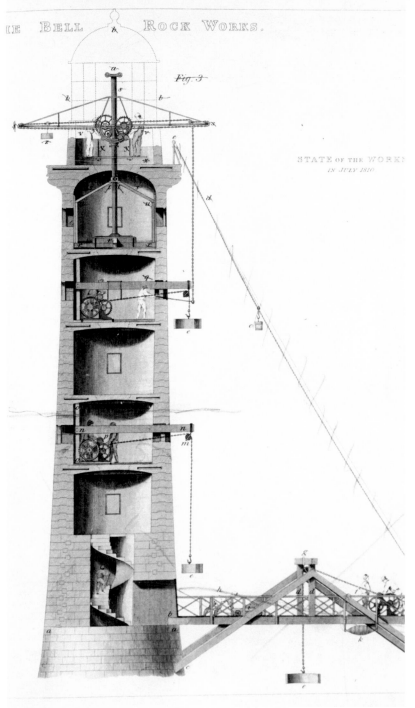

HE BELL ROCK WORKS.

Fig. 3

STATE OF THE WORKS
IN JULY 1810

148 The Crystal Palace

> 44 Extruded Steel Beam **> 46** Open Web Truss **> 60** Barrel Vault

 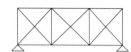

Location: London, UK
Completion date: 1851
Area: 1,848 x 454 ft
Materials: Iron and glass
Engineer: Joseph Paxton

The Crystal Palace was designed to house 100,000 exhibits for the Great Exhibition of 1851. The first international exposition of its kind, it was conceived by Queen Victoria's consort, Prince Albert, in order to celebrate technology from around the world at the height of the Industrial Revolution.

Paxton's design was based on his experience of designing greenhouses, which were in turn said to be influenced by the structure of the Amazonian water lily – radial ribs, strengthened by slender cross-ribs. These principles produced a lightweight structure that was also capable of using prefabricated components – the first of its kind. It was made from "ferro-vitreous" iron, timber and glass, and the dimensions of the structure were based on the largest sheet of glass that could be manufactured at the time.

One of the major advantages of Paxton's ferro-vitreous iron-and-glass design was the building's extreme simplicity – all the principal elements of the building were arranged in multiples and sub-multiples of 24 ft. The total height from the ground floor to the top of the barrel-vaulted transept roof was 108 ft. The transept was added to the design to help enclose some 90 ft-tall elm trees. In total 900,000 sq ft of glass was used, and the entire structure was demounted from its first site in Hyde Park and reconstructed in south London.

The heights of the columns ranged from 16 ft 9 in to 18 ft 4½ in, with the taller versions on the ground floor. Each column was formed of four flat and four cylindrical faces, and the external diameter of all the columns was 8 in on their square face. The base of each column was fixed to a flat plate 2 ft in length and 1 ft in width. This in turn was fixed to a mass of concrete with a length and width of 2 ft by 3 ft and varying in depth from 1 ft to 4 ft.

There were nine varieties of girders and trusses, each 3 ft in depth. There were three different strengths of 24 ft cast-iron girders, one strength of 24 ft wrought-iron trusses (wrought iron was used for its tensile strength), one strength of 48 ft wrought-iron trusses and three strengths of 72 ft wrought-iron trusses.

1 Floor plan of the Crystal Palace.
2 View of the Crystal Palace, *c.* 1930.

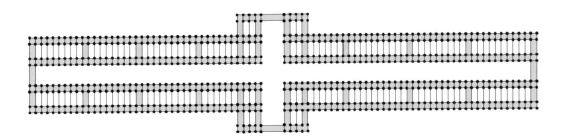

150 Clifton Suspension Bridge

> **46** Suspension

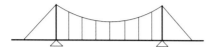

Location: Bristol, UK
Completion date: 1864
Length: 702 ft
Materials: Masonry and steel
Engineer: Isambard Kingdom Brunel

When constructed, the Clifton suspension bridge was the longest single-span bridge in the world. It spans 702 ft, measured from the center of each pier, over the 250 ft-deep Avon Gorge. Although Brunel was 24 when he won the competition to design the bridge, it was only completed five years after his death in 1859. Brunel's colleagues in the Institution of Civil Engineers felt that completion of the bridge would be a fitting memorial; in 1860, Brunel's Hungerford suspension bridge over the Thames in London was demolished to make way for a new railway bridge and its chains were purchased for use at Clifton.

Brunel was born in 1806, and by the 1820s was working with his father on the early stages of the construction of a tunnel under the Thames. In 1833, he was appointed chief engineer of the new Great Western Railway and went on to create innovative designs for everything from tunnels, railways and bridges to harbors, prefabricated buildings and ships. Brunel was always on the lookout for new technologies (in particular expanding the use of iron) and revolutionized society's approach to both mechanical and structural engineering.

The piers (towers) of the Clifton suspension bridge are built principally of local Pennant stone. The 86 ft high tower on the Leigh Woods side of the Avon Gorge stands on an abutment that was originally thought to have been solid but has since been found to consist of a series of 12 vaulted chambers linked by small tunnels. The wrought-iron chains from which the deck is suspended dip to 70 ft and are anchored back into the rock at either end through tunnels that are 60 ft below ground level. Roller-mounted "saddles" are used at the top of each tower to absorb the forces created by the movement of the chains when under load.

The road deck itself is carried by two, huge girders (assembled in sections each of which was connected to the chains using vertical suspension rods) and is made from 5 in thick pine beams (known as sleepers) with 2 in thick floor planks. In total, the bridge used 1,600 tons of steel.

The bridge was designed to carry horse-drawn traffic but now carries 11,000 to 12,000 motor vehicles every day.

1 The Clifton suspension bridge, Avon Gorge.
2 Detail of connections between the deck and the wrought-iron chains. The road deck is suspended from the chains by 81 wrought-iron rods on each side, which range from 65 ft in length at the ends to 3 ft in the center.

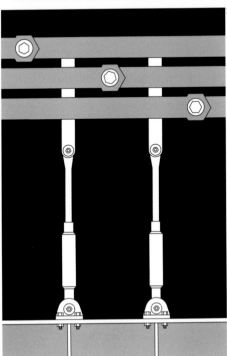

152 Tropical House (Pavilion Demontable)

> **58** Portal Frame　　> **72** Thermal Comfort　　> **84** Air Movement

Location: Brazzaville, Congo, Africa
Completion date: 1951
Area/Weight: 1,600 sq ft/11 tons
Materials: Sheet steel, aluminum
Engineer: Jean Prouvé

Jean Prouvé was a trained locksmith who pioneered design and fabrication processes for furniture and modular construction. His prototype lightweight, prefabricated metal building system represented a major step in the industrialization of architecture.

Fabricated in Prouvé's French workshops, the components for two modular buildings, designed in collaboration with his brother Henri, were completed in 1951 and flown to Africa in the cargo hold of an airplane. The houses are designed on a simple, 3 ft 3 in, grid system and consist of a series of ground beams connected to fork-shaped portal-frame supports of folded sheet steel. Prouvé's unique portal frame system provides horizontal bracing to the entire structure and support for the roof and walls, enabling the wall panels to be kept as light as possible. All but the largest structural elements are aluminum, and no piece is longer than 13 ft – which corresponds to the capacity of the brake press (folding machine) – or heavier than 200 lb, for easy handling by two men.

The fabrication of the portal elements and other structural components was, and remains, an innovation in the construction industry, utilizing sheet-metal techniques of folding, pressing and profiling to get maximum strength out of a light and flexible substrate. In techniques more often employed in the automotive and aeronautical industries, Prouvé pioneered a prefabricated, component-based approach to construction that remains highly influential within the profession and can clearly be seen in the English hi-tech movement of the late twentieth century (Norman Foster, Richard Rogers *et al*). Interestingly Jean Prouvé was one of the jury members on the judging panel for the Centre Pompidou, Paris, won by Renzo Piano and Richard Rogers. The structural integrity of Prouvé's system meant that the building could be propped up or supported by a series of designated columns locally prepared in advance of their arrival, which deals with both ventilation and potential damp. Again, this technology is more akin to that of a vehicle, be it a plane, train or automobile and shares the technological ambition of Buckminster Fuller's Dymaxion Dwelling Machine (or Wichita House) of 1946.

To cope with the extremes of the tropical climate, the outer light-reflecting skin, consisting of adjustable brises-soleils of each building that shielded the structures from direct sunlight, was separated from the inner insulated skin of sliding doors and fixed panels. The buildings used self-shading natural cooling and ventilation with the profile of the roof "wings" designed to exploit the stack effect, thus cooling the structures.

In 2001, one of the buildings (pockmarked by bullet holes) was brought back to Paris, and was restored and reassembled. It has subsequently been moved to the USA where it was put on sale by Christie's, as part of a contemporary design auction and sold for the not inconsiderable sum of US$6 million. The eventual economic value of this highly prized prefab could not have been envisaged by Prouvé, but the ease of relocation most definitely was.

1 Rotating louvre used in the construction of the Tropical House.
2, 3 Interior views of the Tropical House as reconstructed.

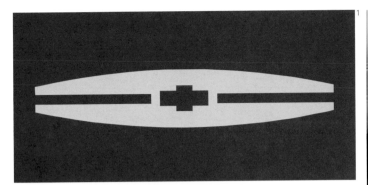

154 Kresge Auditorium

> **50** Reinforced-Concrete Slab

> **98** Acoustics

**Location: MIT campus,
Massachusetts, USA
Completion date: 1955
Height: 50 ft
Span: 160 ft
Materials: Reinforced concrete
Architect/Engineers/Acousticians:
Eero Saarinen, Ammann & Whitney,
Acentech**

Eero Saarinen designed the Kresge auditorium together with the MIT chapel; the two buildings are separated by a green space. The ensemble is recognized as one of the best examples of mid twentieth-century modern architecture in the USA. This was Saarinen's first major public project, but despite its architectural success there was a perception that it was an unsuccessful piece of civic design. Saarinen himself later remarked that the two inward-orientated structures were "too egotistical... theoretically, it is a very graceful building. Structurally, it's quite a rational building. But, if you look at it, isn't it a little bit too earthbound? ... It did not have the soaring quality or sense of lightness that one wanted."

The reinforced-concrete shell structure takes the form of a one-eighth segment of a sphere, and is supported at three points by concrete and steel abutments. Triangular in plan, the roof is clad in copper, and steel-framed glass walls complete the climatic enclosure. Although the first double-curving, lattice structure (grid-shell) had been built out of steel sections as far back as 1897 (by V.G. Shukhov), thin, reinforced-concrete shell structures were still largely experimental at the time, and the roof weighed only 1,200 tons. The Kresge auditorium was the first (thin) concrete shell of this size in the USA.

The dome touches the ground at three points almost 150 ft apart and is only 3½ in thick at its uppermost point.

The building was designed to be an auditorium with seating for 1,226 people. As it is a column-free space, each person has an unobstructed view of the stage area. The acoustic quality of the space was controlled by the innovative use of free-hanging ceiling baffles that could absorb and direct sound. In 1998, as part of a general refurbishment, the acoustic quality of the space was further enhanced by new acoustic panels that were designed to reflect some of the sound back to the stage and spread the rest as uniformly as possible.

There has been much speculation that the design of Saarinen's subsequent TWA Terminal Building at Kennedy International Airport, New York, may have been influenced by Jørn Utzon's winning entry for the Sydney Opera House competition, for which Saarinen was a jury member (see page 166). Conversely, it might be that Utzon was influenced by the then constructed shell of the Kresge auditorium, completed in 1955.

1 Diagram showing the one-eighth segment of a sphere that makes up the shell structure.
2 View of the auditorium.

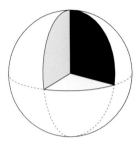

156 Palazzetto Dello Sport

> **40** Reinforced-Concrete Column   > **48** Profiled Plate: Reinforced Concrete   > **62** Lamella Dome

Location: Rome, Italy
Completion date: 1959
Materials: Reinforced concrete –
ferrocement
Dimensions: diameter 194 ft,
height 69 ft
Engineer: Pier Luigi Nervi

Known as a lamella dome, the palazetto was built from reinforced concrete for the 1960 Olympic Games. The 5,000-seat "little palace" (designed with Annibale Vitelozzi) has prefabricated diamond-shaped sections that fit together to form transversing helical ribs. The dome springs from a series of exposed, prefabricated Y-shaped piers sloped to receive their diagonal thrusts. These piers allow for large openings at ground level.

Nervi was an engineer and building contractor who became internationally famous for his invention of ferrocement – a dense concrete heavily reinforced with steel mesh. Between 1935 and 1942, he designed a series of aircraft hangars for the Italian air force, the first of which at Orvieto (1938) determined their form: long, pointed barrel vaults, constructed on latticed grids, rising from complex triangulated edge-beams. The various-sized ribs were prefabricated from reinforced concrete rather than being poured in place. Instead of using traditional methods for reinforcing concrete, Nervi designed a more balanced, composite material consisting of layers of steel mesh grouted together with concrete. He also pioneered the use of prefabricated concrete panels, thus avoiding costly and time-consuming timber formwork.

In addition to designing large-span vaults and domes, Nervi succeeded in building a sailing boat with a ferrocement hull only ½ in thick.

The Palazzetto Dello Sport is still in regular use, mainly for Basketball. Also in

Rome's EUR (Esposizione Universale Roma) district is the much larger, Palazzo Dello Sport, also by Nervi.

Pier Luigi Nervi won the Royal Institute of British Architects (RIBA) Gold Medal in 1960.

1 Cross-sectional drawing of the Palazzetto dello Sport, Rome.
2 Exterior and interior views of the Palazzetto Dello Sport.

158 The Houston Astrodome

> **46** Open Web Truss

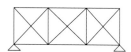

> **62** Diamatic Dome

> **84** Air Movement

Location: Houston, Texas, USA
Completion date: 1965
Height: 208 ft
Weight: In excess of 2,700 tons
Materials: Concrete and steel frames
Architects: Hermon Lloyd & W.B. Morgan
and Wilson, Morris, Crain & Anderson
Structural Engineers: Walter P. Moore
Engineering Consultants

The Astrodome was the first ballpark/ stadium to have a roof over the playing field. The dome has a clear span of 642 ft and is constructed with trussed ribs that are 5 ft deep. The roof, which covers an area of 350,000 sq ft, is held in place by a 414 ton tension ring. An 18-story building would fit inside the Astrodome.

The stadium seats up to 66,000 people, and the central air-conditioning system has to circulate 2.5m cu ft of air a minute. The Astrodome was also designed with an astonishing 28,000 car parking spaces over a site of 280 acres.

The Houston Astrodome marked what many saw as the birth of modern stadium design. It represented the first totally enclosed multi-purpose stadium in the world. In three configurations the Astrodome can seat 66,000 spectators for concerts, 52,000 for American football and 45,000 for baseball.

When the Astrodome opened the playing surface was natural Bermuda grass. The dome's ceiling contained numerous semi-transparent plastic panels made of Lucite. Players soon complained that glare coming off the panels made it impossible for them to track "fly balls," so all the panels were painted over – which solved the glare problem but caused the grass to die from lack of sunlight. This led to the development of a new artificial playing surface that used a synthetic woven "grass" matting, which was subsequently named Astroturf after its originating venue. The Astrodome was the first stadium in North America to use

separate Astroturf pitches for baseball and football, each housed in a storage pit in the center of the field and rolled out on a cushion of air. Interestingly, the development of this artificial playing surface only increased the "multi-purpose" potential of such a large arena for exhibitions and conferences as well as sporting events. The Astrodome also led the way in the development and use of large-scale electronic displays, and in well-serviced refreshment and washroom facilities that were previously lacking in such large places of congregation.

1 The Houston Astrodome under construction.

160 The German Pavilion

> **66** Mast-Supported Membrane

Location: Montreal, Canada
Completion date: for Expo '67
Size: Max. length 426 ft, max. width 344 ft
Covered area: 86,111 sq ft
Mast heights: 46 to 125 ft
Architects: Frei Otto, Rolf Gutbrod
Engineer: Fritz Leonhardt

As one of the world's leading authorities on lightweight tensile membrane structures, Frei Otto has pioneered advances in structural mathematics and civil engineering. Otto's career bears a similarity to Buckminster Fuller's: both were concerned with space frames and structural efficiency, and both experimented with inflatable buildings. In the 1950s, Otto used models to define and test complex tensile shapes. As the scale of his projects increased, he pioneered a computer-based procedure for determining their shape and behavior.

"Frei Otto and Rolf Gutbrod attempted, with this competition-winning project, to create a man-made landscape. The cavernous interior contained modular steel platforms arranged at different levels. The entire area was covered by a single membrane of irregular plan and varying heights. Its contours were determined by the high points of the masts and the low points where the membrane was drawn, funnel-like, down to the ground. Eye loops filled with clear plastic material accentuated these points and the saddle surfaces they created. The prestressed membrane consisted of a translucent plastic skin hung from a steel wire net, which, by eye, ridge, and edge ropes, was connected with the mast heads and anchor blocks." From Ludwig Glaser, *The Work of Frei Otto*, page 109.

This cable net structure was supported by ½ in twisted steel cables on an 18 in grid (so that it could be climbed on), with 2 in cables employed at the edges and on the ridge. The translucent fabric skin was a woven flame-resistant polyester substrate.

The pavilion also employed large sweeping canopies of acrylic glass, stabilized by steel cables that were used for the first time on such a scale.

Originally built only for the duration of the Expo, the structure was retained by Montreal's authorities for the young people of the city until it was finally dismantled in 1976.

The Expo cable net also housed two timber lath shell structures for use as auditorium and foyer areas. These were early antecedents of the Downland Gridshell by Edward Cullinan (see page 60).

1 Diagrams showing the contours of the tensile membrane covering the pavilion
2 The German Pavilion, Montreal, 1967, enclosed an 86,111 sq ft area.

162 The USA Pavilion

> 62 Geodesic Dome   **> 72** Thermal Comfort

Location: Montreal, Canada
Completion date: for Expo '67
Height: 200 ft
Diameter: 250 ft
Materials: Tubular steel
Architects: R. Buckminster Fuller,
Shoji Sadao

Buckminster Fuller invented the geodesic dome, and described it, in its 1954 patent application, as a "spherical mast" distributing tension and compression throughout the structure, which in this example constituted around three quarters of a sphere. The pavilion was constructed for the Montreal Expo, and consisted of a double-layer space grid that employed three-dimensional modules, triangular on the outside, hexagonal on the inside. Connecting them together in the shape of a dome distributed the structure's weight over the whole surface. Fuller's geodesic dome was originally weathered using 1,900 molded acrylic panels, which incorporated six triangular sunblinds within each six-sided panel, and were automatically opened or closed in response to the relative movement of the sun in relation to the structure; an early example of a responsive or "intelligent" computer-controlled shading system. In 1976 fire destroyed the acrylic cladding, but this remarkable structure still exists as a museum overlooking the city of Montreal.

On the occasion of Fuller receiving the coveted American Institute of Architects (AIA) Gold Medal in 1970, the AIA described the geodesic dome as "the strongest, lightest, and most efficient means of enclosing space yet known to man."

Interestingly, the structure is not entirely geodesic. If you look for the equator (or the horizontal half-point) you will see that the horizontals below (towards the ground) are parallel and of decreasing circumference, whereas the structural geometry above the equator is purely geodesic.

The Montreal dome is not the biggest geodesic ever constructed; it was surpassed by Donald Richter's 415 ft hangar built to enclose Howard Hughes' seaplane *Spruce Goose* in 1982: Richter had been a student and assistant of Fuller. More recently, the Buckminster Fuller Institute listed the Fantasy Entertainment Complex, Kyosho, Japan, as currently the world's largest geodesic structure with a diameter of 710 ft. Fuller had proposed in 1950 a dome that would enclose the whole of Manhattan in a 2 mile diameter geodesic geometry whose physical enclosure would have weighed significantly less than the volume of air contained inside.

In 1997, Harold Kroto, Richard Smalley and Robert Curl were awarded the Nobel Prize for Chemistry, for their discovery of a new class of carbon. Both Kroto and Curl are said to have visited the Montreal dome and the new carbon molecule was subsequently named (on their insistence) Buckminster-fullerene, or fullerene for short.

As well as the USA Pavilion, Expo '67, also featured Frei Otto's German Pavilion (see page 160) and Moshe Safdie's Habitat, a unique development of 158 residences from 365 prefabricated concrete construction modules.

In 1990, Environment Canada purchased the dome site and transformed the USA Pavilion into Biosphère, a museum designed to raise awareness of the St Lawrence River and Great Lakes ecosystem.

1 The USA Pavilion for Expo '67, Montreal, was the world's largest geodesic dome when it was built. Photographed in 2004.
2 Detail of the pavilion's geodesic tubular steel structure with its twelve- and six-way connecting joints. Photographed in 2004.
3 A Buckminster Fuller patent drawing, which illustrates different geodesic configurations based on the "great circle" subdivision of a sphere.

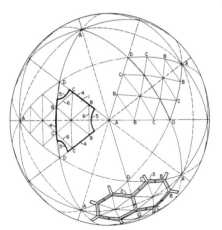

¹⁶⁴ Self-Build House

> **42** Pad Foundation   > **58** Post and Beam

Materials: Timber
Architect: Walter Segal

During the 1970s, the Swiss architect Walter Segal developed a system for building timber-framed housing that was fast, cheap, efficient, and simple. Segal's achievement was to devise a way of simplifying the process of building so that it could be undertaken by anyone, cheaply, and quickly. He insisted that his was an approach, not a system, and he made no claims for originality or patents. He considered that a house, rather than being a fully finished product right from the start, is a simple basic structure that grows over time as needs grow and as labour and income can be spared.

The framing was based on modular dimensions to avoid waste and to facilitate alterations and enlargement. Specialist "wet" trades such as plastering and bricklaying were replaced by lightweight cladding, lining, and insulation. Segal's purpose was that someone with limited knowledge of construction could build themselves a house using nothing more than a manual and construction elements that were readily available from local suppliers – an idea known today as self-build.

In the late 1970s, Lewisham Borough Council in London believed the Segal method would be perfect for the development of a number of steep, soft-soiled council plots unsuited to conventional building.

The plots were duly offered to people on the council's waiting list and they set about building their own homes (1977–82) on a site that was later named Segal Close. A second development, Walter's Way, followed in 1985–87.

The modular grid was determined by the sizes of standard building components, to maximize efficiency and minimize waste. Materials were chosen on the basis of ease of use in terms of size, weight, and their ability to be manipulated using simple power tools. Foundations and groundworks were reduced to a minimum by using a series of discrete pads, and components were assembled using "dry" joints formed with bolts and screws. The buildings are adaptable and easily extended, and the method can produce both one- and two-story buildings with either flat or pitched roofs, and can include double-height spaces.

1 Segal Self-Build. Diagram showing the various elements and the sequence in which tasks are done.
2 Walter's Way, London, 1985–87.

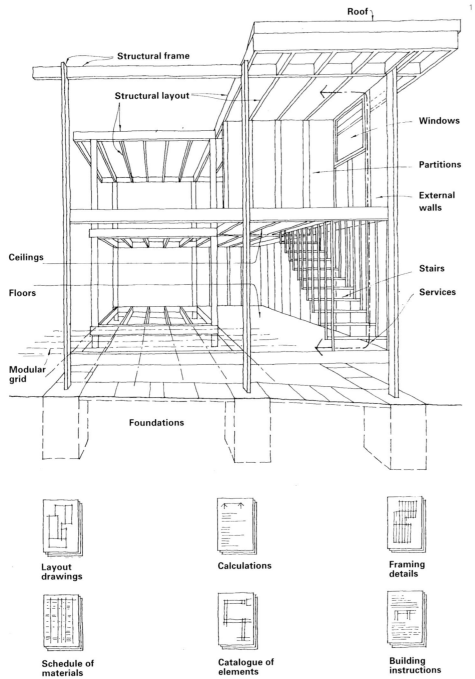

Roof

Structural frame

Structural layout

Windows

Partitions

External walls

Ceilings

Floors

Stairs

Services

Modular grid

Foundations

Layout drawings

Calculations

Framing details

Schedule of materials

Catalogue of elements

Building instructions

¹⁶⁶ Sydney Opera House

> **50** Reinforced-Concrete Slab   > **60** Pointed Vault   > **98** Acoustics

Location: Sydney, Australia
Completion date: 1973
Size: 610 x 400 ft
Materials: Reinforced concrete
Architect: Jørn Utzon
Engineer: Ove Arup

Jørn Utzon's winning entry in an international architectural competition in 1957 was said to have been excluded by the technical judging panel, but later reinstated on the recommendation of one of the judges, architect Eero Saarinen.

Work on the Sydney Opera House started in 1959. Geometrically, each half of each shell is a segment of a sphere; however, the "sails" were originally designed as parabolas, for which an engineering solution could not be found. Although described as reinforced-concrete shells, each one employs a series of hollow concrete ribs that support a total of 2,194 precast-concrete roof panels (or distinctive arrow-shaped tile lids) which are in turn clad with a total of over 1m tiles. The tiled surface is highly detailed and uses two types of tile – one white, one cream – with clearly expressed joints. It was the intention of the architect that the building operate simultaneously at both building and human scale.

The opera house is supported on 580 concrete piers sunk up to 82 ft below sea level.

The design of the shells involved one of the earliest uses of computer analysis in order to understand the complex forces that the shells would be subject to, and it took some years to find the solution – that all the shells would be created as sections from a sphere and be supported on arched ribs. This solution avoided the need for expensive formwork construction by allowing the use of precast units which could be tiled at ground level. Large parts of the site were used (throughout construction) as a factory for these precast components.

The design and construction process was high profile and not without controversy. The opera house was to be built on Bennelong Point adjacent to Jon Bradfield's Sydney Harbour Bridge and would be visible from all sides.

In 1966 during an interview for Danish television, Utzon said "...it was an ideal project for an architect...first because there was a beautiful site with a good view, and second there was no detailed programme."

The job proved a highly complex one, even with the assistance of engineering firm Ove Arup and, more specifically Jack Zunz (see HSBC Headquarters, page 168) and young assistant Peter Rice (see Pavilion of the Future, page 170).

On 1 March 1966 Utzon announced that he was quitting the job, a forced resignation because he had not received fees owing to him and there had also been a more general breakdown in communications with the client.

On 20 October 1973 the Sydney Opera House (completed by Hall, Todd and Littlemore) was officially inaugurated by Elizabeth II. Utzon never returned to visit the opera house finished by others, but in 1999 he was re-engaged to work on the building's interiors, assisted by his son and partner, Jan, who oversaw the work. The subsequent renovation of the reception hall led to its being renamed the Utzon Room.

1 Diagram shows the segments of a sphere that make up the "sails" of the Sydney Opera House.
2 Exterior photograph showing the tile detailing.

168 HSBC Headquarters

> 34 Open Web Truss Column   **> 46** Bowstring Truss   **> 72** Thermal Comfort

 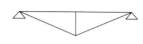

Location: Hong Kong, China
Completion date: 1985
Height/Weight: 600 ft/
33,000 tons of steel
Architect: Foster + Partners
Engineer: Arup (Jack Zunz)

A modular building with many prefabricated elements, the Hong Kong and Shanghai Bank headquarters employs just four vertical ladder-type, Vierendeel trusses as columns from which a series of double-cantilever elements (suspension trusses) are used to suspend the floors in groups of seven. The two-story suspension truss frames divide the structure vertically into five visually separated components with the two-story trusses inhabited as reception, dining and conference areas. These are connected to open-air terraces, cleverly created by pulling back the climatic envelope away from the structural frame. The terraces are designated recreational spaces, but are also treated as refuge zones in case of fire. The cross-bracing at the suspension truss levels is pin-jointed at the masts, a technology more commonly associated with civil engineering structures such as bridges, but used to cope with the huge loads and avoid bulky and labour-intensive bolted connections.

The unique "exoskeletal" structure provides large internal spaces that are clear of supports and allow for maximum future flexibility of the office space, of which this building provides more than 1m sq ft.

The HSBC building has 47 storys and four basement levels, and is clad in 4,500 tons of aluminum.

Natural sunlight is reflected into the lower-level internal spaces through the use of two sun scoops; one internal, one external. The external sun scoop consists of a giant bank of movable mirrors mounted on a truss structure on the South side of Level 12 (to correspond to the top of the 170 ft high atrium). These heliostats are programmed to track the sun throughout the year and reflect sunlight onto the internal sun scoop (or fixed bank of convex aluminum mirrors) which directs light around the atrium towards the plaza below. In addition, sun shades are provided on the external façades to block direct sunlight entering the building and thus reduce heat gain.

All flooring is made from lightweight movable panels in order to gain access to the service infrastructure below, including air-conditioning, electrical and telecommunication points.

Norman Foster had at the time never built a building of more than four storys but successfully employed a unique structural system to create what ultimately became home to 4,000 bank workers.

The design and construction of the project produced over 120,000 drawings and was a truly international collaboration with cladding from the USA, structural steelwork from the UK and suppliers and fabricators and contractors from Hong Kong and Japan.

At the time of the HSBC headquarter's completion in 1985 it was the most expensive high-rise building ever constructed, costing an estimated US$670 million. The project propelled Norman Foster into global recognition and allowed for the closest that architecture has to research: a significant budget with the means to indulge, prototype and test architectural technology to its contemporary limit.

1 Diagram of a two-story suspension truss.
2 The Hong Kong and Shanghai Bank headquarters.

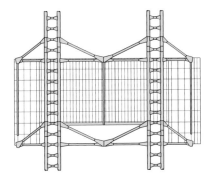

170 The Pavilion of the Future

> **36** Concrete Blocks

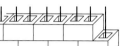

> **46** Bowstring Truss

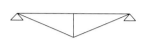

> **60** Barrel Vault

Location: Seville, Spain
Completion date: For Expo '92
Height: 127 ft
Materials: Stone and steel
Architects: Martorell Bohigas Mackay (MBM)
Engineer: Peter Rice

Peter Rice was born in Ireland in 1935. He joined Arup in 1956 and much of his early career was spent working on the Sydney Opera House, contributing to geometric and model studies and carrying out analytical studies. Following his time in Australia he took a year's sabbatical as visiting scholar at Cornell University, where he wanted "to study the application of pure mathematics to engineering problems. I think that a more thorough understanding of the nature of the equations used to solve structural problems in design could lead to a better conditioned solution and ultimately to a better choice of structural component."

For the Pavilion of the Future, Rice designed a series of interlocking arches, visually resembling the stone aqueducts of old, that employed the compressive strength of stone and the tensile strength of steel. Unlike in reinforced concrete, however, the steel used to both reinforce and stabilize the structure was expressed externally. Similarly, the stone elements were reduced to the minimum cross-section needed for strength. The stone, Rosa Porina granite, was cut to extremely high tolerances (of less than 1/60 in) by computer-controlled cutting machinery. Each piece was 8 in square and 4 ft 6 in long, and the pieces were joined using epoxy resin to form prefabricated modular units of 16 ft 5 in by 1 ft 8 in by 1 ft 8 in.

Because the weight alone of these "skeletal" stone arches was not enough to provide stability, each arch was vertically braced using a unique system of stainless-steel rods and ties.

One difficulty encountered during construction was that the natural rift plane of the stone (the grain in timber) was not consistent or directionally uniform, so the stones' specific directional strength was hard to calculate.

1 Diagram detailing the interlocking arches of stone and steel.
2 The stone and steel "braced" arches of the Pavilion of the Future. Photographed in 1992.

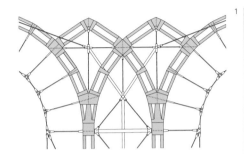

172 The Portugal Pavilion

> **46** Suspension

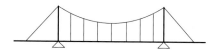

> **50** Reinforced-Concrete Slab

Location: Lisbon, Portugal
Completion date: For Expo '98
Size: 165 x 220 ft
Weight: 1,500 tons
Materials: Reinforced concrete
Architect: Álvaro Siza Vieira
Engineer: Arup

Portugal's most famous living architect, Álvaro Siza, constructed the exhibition building of the host country for Expo '98. The brief for the pavilion was that it was to be the venue for a temporary multimedia presentation for the duration of the Expo and provide a large external area for state visits and other ceremonial occasions. The architectural critic and theorist Kenneth Frampton, writing in *Álvaro Siza – Complete Works*, said this indeterminate brief had led to Siza designing the project as a palace and "Bringing him to adopt a monumental syntax."

The curved reinforced-concrete roof is supported on two ceramic-clad porticoes. The stainless-steel tensile support cables are visible at each end of the span where they are attached to loadbearing walls. These walls sit in the same plane as the reinforcing bars and are designed to withstand the enormous lateral loads, like the masts and tie-backs of a suspension bridge. Engineered by an Arup team headed by the celebrated Cecil Balmond, the canopy's structural support was separated from the other parts of the building due to Lisbon's seismic activity.

The roof, 8 in of concrete, drops only 10 ft over its 65 m span. Devoid of lighting or drainage systems, this thin concrete skin spans the 213 x 164 ft plaza like a sail. Rainwater runs down the slightly sloping roof towards the harbour basin and falls freely to the ground.

This huge "concrete catenary" held between two free-standing façades bears striking similarities to German engineer Fritz Leonhardt's swimming hall at Wuppertal (1954–56), which was a 213 ft wide concrete catenary held between the upper edges of two grandstands. Leonhardt, incidentally, worked with Gutbrod and Otto on the German Pavilion for the Montreal Expo (see page 160).

1 Diagram shows the ceramic-clad porticoes supporting the curved concrete roof.
2 The Portugal Pavilion pictured during Expo '98.

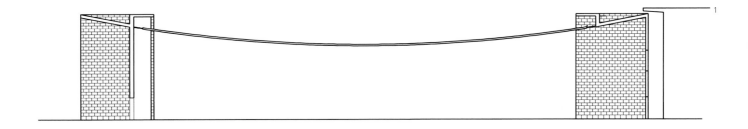

1

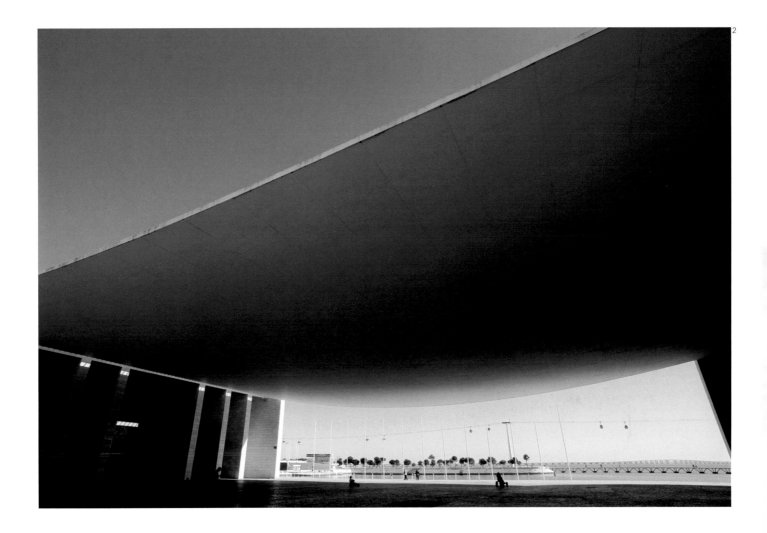

2

174 The Lord's Media Centre

> **50** Coffer   > **60** Ribbed Dome   > **72** Thermal Comfort

Location: London, UK
Completion date: 1999
Capacity: 250 people
Weight: 110 tons
Materials: Aluminum
Architects: Future Systems
Engineer: Arup
Contractor: Pendennis Shipyard

The Media Centre for Lord's cricket ground in north London was designed to house 120 writers, 100 broadcasters and a 50-seat restaurant.

The main building is prefabricated in 26 sections approximately 10 by 65 ft, each weighing between 4.4 and 6.6 tons. The shell was preassembled off site in a boatbuilder's yard and then reassembled at Lord's. The building industry refused to tender for the building and only one boatbuilder took up the challenge.

Amanda Levete, a director of Future Systems, says that when the practice was invited to submit ideas for a new Lord's Media Centre, "we decided to go for broke." Critic Martin Pawley, writing in *The Architects Journal* in 1998, described the Media Centre as "a semi-mythical beast, half boat, half gazing eye, which has at last become believable."

Monocoque structures (also known as stressed-skin structures) are lightweight and are commonly used in the aircraft, automobile and shipbuilding industries. The Media Centre is the first all-aluminum semi-monocoque building in the world, and is composed of an aluminum skin supported by a ribbed (coffered) framework of ¼ in and ½ in aluminum sheet metal. The prefabricated sections were welded together on site, with the welds subsequently ground down and spray-painted, creating a continuous smooth surface. The monocoque structure allows for unobstructed views onto the pitch.

The Media Centre is carried 49 ft above ground on two concrete towers that contain stairs and lifts, and from which the rigid aluminum structure cantilevers in every direction. The structure also contains an inclined non-reflective glass façade and air cooling to mitigate the effect of direct sun.

At the specific request of the British Broadcasting Corporation's (BBC's) venerable radio commentary team on *Test Match Special*, there is one opening window in the otherwise seamless glass façade. This window allows the commentators to hear the ball being hit without the aid of artificial amplification.

1 The prefabricated sections of the Media Centre, engineered in Europe and welded together on site.
2 Rear view of the Media Centre showing the two concrete support towers.
3 On site assembly of one of the prefabricated aluminum sections.

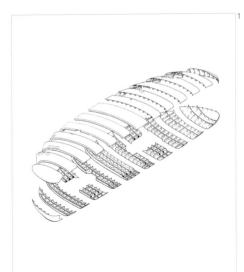

176 30 St Mary Axe

> 62 Lamella Dome   **> 72** Thermal Comfort   **> 84** Air Movement

Location: London, UK
Completion date: 2004
Weight: 77,000 tons
Height: 590 ft
Materials: Glass and steel
Architects: Foster + Partners
Engineers: Arup
Contractor: Skanska UK
Steelwork fabrication: Hollandia BV,
Viktor Buyck Steel Construction NV

Most tall buildings get their lateral stability from a rigid central core. However, with this tower the exoskeletal frame braces the structure through triangulation as well as carrying the floor loads – each floor rotates 5 degrees from the last. Despite its overall curved glass shape, there is only one piece of curved glass on the building – the lens-shaped cap at the very top – tapering into the crown, it reduces wind resistance.

Windows in the light wells open automatically to augment the air-conditioning system with natural ventilation. The atria are arranged in a spiral so that air drawn into the tower via the light wells circulates around the building due to the differences in external air pressure. The office areas consist of a double-glazed outer layer and a single-glazed inner screen, which sandwich a central ventilated cavity that contains solar-control blinds.

The building contains 333 piles in the foundations, 22 miles of structural steel at 12,000 tons, and 258,300 sq ft of glass cladding. The lifts can carry a maximum of 378 people at any one time reaching speeds of 19 ft per second, and almost all the plant space is located externally on an adjacent six-story structure at 20 Bury Street.

This 41-story office block provides 822,370 sq ft of accommodation, including offices, a new public plaza at ground level, and a club room atop the structure that provides a 360-degree panorama of the city. 30 St Mary Axe is claimed to be London's first environmentally sustainable high-rise, utilizing up to 50 per cent less energy to drive its heating and cooling system.

The shape of the tower was specifically tailored to its immediate local environment and the constraints of a tight City of London site. Its geometric profile was designed to reduce reflections, thus making the tower appear more transparent. The shape of the tower is also designed to reduce the effects of "down-draft" or winds at ground level that occur with high-rise developments, while creating pressure differentials across the service area to drive the unique natural ventilation system.

In 2004, 30 St Mary Axe was awarded the Royal Institute of British Architects (RIBA) Stirling Prize for building of the year. In his acceptance speech Norman (Lord) Foster said that the structure "... is an embodiment of the core values that we have championed for more than thirty years: values about humanizing the workplace, conserving energy, democratizing the way people communicate within a building, and the way that the building relates to the urban realm."

This project develops ideas explored at the HSBC headquarters in Hong Kong (1985), the Commerzbank in Frankfurt am Main (1997) and the theoretical Climatroffice, an early collaboration between Foster and Buckminster Fuller (1971), which explored a dynamic mix of nature, office and space frames as future workspaces.

1 Segment revealing the internal structure of the tower.
2 30 St Mary Axe, or the "Gherkin" as it is more popularly known.

178 "D" Tower

> **38** Loadbearing Laminate Wall   > **56** Portal Frame (curvilinear plan)

Location: Doetinchem, the Netherlands
Completion date: 2003
Height: 40 ft
Materials: Glass-reinforced epoxy (GRE)
Architects: Nox Architecture

The "D" Tower is made from 19 separate, laminated panels, assembled on site, and made from glass-reinforced epoxy (GRE) that was cast in molds. The molds were assembled from expanded polystyrene blocks that had been cut to form the complex surfaces by computer numerically controlled (CNC) milling machines.

The structural geometry is very similar to that of a Gothic vault, where columns and surface share the same continuum, except that the parts are glued and bolted together.

The reason for the use of epoxy is that the building was designed to glow, using different-colored lamps that responded to the mood of the inhabitants of Doetinchem (who were able to send messages to the building via a web site). The reason for the complex geometry is that the tower was designed to simulate a large, beating human heart.

1 Assembly schematic for the 19 components.
2 Cutting the expanded polystyrene blocks.
3 Each mold is built using an assembly of blocks; the panel is formed by laminating the mold.
4 Test assembly.
5, 6 Night-time views indicating the "chromatic" mood swings.
7 "D" Tower.

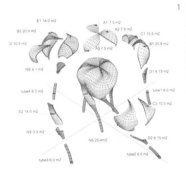

1

5

7

2

6

3

4

180 Davies Alpine House

> **60** Barrel Vault   > **72** Thermal Comfort   > **84** Air Movement

**Location: Royal Botanic Gardens, Kew,
London, UK**
Completion date: 2005
Height/width/span: 33 ft/49 ft/62 ft
**Materials: Painted steel and
structural glass**
Architects: Wilkinson Eyre
**Engineers: Structural – Dewhurst
MacFarlane, Environmental – Atelier 10**

Each twin arch is formed from an upper and
lower arch, spanned by tension rods which
are anchored to a concrete retaining wall at
ground level. The arches were fabricated
from mild steel flat plates cut to a curve and
welded together to make rectangular
sections measuring 9½ x 4¾ in. The stainless-
steel tension rods also carry the faceted glass
panels. Alpine plants have to be dry, cool and
well ventilated, and need to receive lots of
light. The height of the glasshouse
encourages air to move upwards (through
vents at top and bottom) and the concrete
base acts as a heat sink – a small fan draws
cool, night-time air through a maze of
concrete passages beneath the floor, and a
series of vents directs the cool air upwards
during the day (see also Thermal Mass,
page 80). The building is north-south oriented
so as to present a narrow profile to the sun.
Low-iron glass gives nearly 90 per cent
transparency, while mechanical solar shading
may be deployed across the façade like a fan.

  1 "Exploded" axonometric drawing
describing the structural logic and
principal structural and cladding
elements.
  2 Thermal control. Cross-section
drawing showing air movement and
temperature.
  3 View of the Davies Alpine House
with the solar shading deployed.

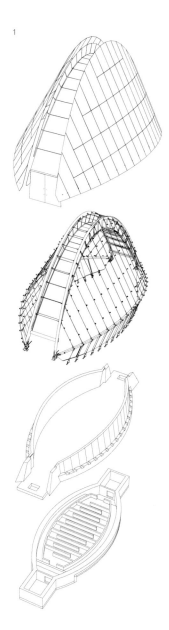

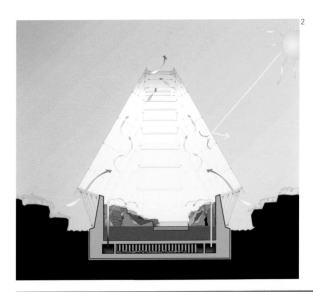

182 International Space Station

Location: Stratosphere
Start date: 1998
Engineer(s): International
Weight: 471,444 lb*
Habitable Volume: 15,000 cu ft*
Dimensions: Span of Solar Arrays 240 ft*
Length: 146 ft from Destiny Laboratory
to Zvezda; 171 ft with a Progress
vehicle docked
Truss: 191 ft
Height: 90 ft

The future of construction technology can perhaps be illustrated by the orbital assembly of the international space station (ISS). Pre-fabricated construction elements are ferried into space and assembled using both manpower and sophisticated robotics. The construction elements themselves are the product of an advanced materials science that results in increased lightness of elements along with their ability to withstand extreme environmental conditions (though not gravity!). The space shuttle and two types of Russian launch vehicles have so far launched a total of 45 assembly missions.

*Statistics as of December 2006

1 ISS Assembly Mission 1: The Zarya control module was launched atop a Russian proton rocket from Baikonour Cosmodrome, Kazakhstan on 20 November 1998. Zarya provides battery power, fuel storage and rendezvous and docking capability for Soyuz and Progress vehicles.

2 ISS Assembly Mission 2A: Space shuttle Endeavour delivered the unity node with two pressurized mating adapters. On 6 December 1998, the STS-88 crew captured the Zarya control module and mated it with the unity node inside the shuttle's payload bay. On Sunday 13 December, space shuttle Endeavour undocked from the young international space station for the return to Earth.

3 ISS Assembly Mission 2A: The STS-101 crew readied the international space station for the arrival of the Zvezda service module. Four new batteries, ten new smoke detectors and four new cooling fans were installed in the Zarya control module. Handrails were installed on the unity node for future space walks.

4 ISS Assembly Mission 4A: The STS-97 crew delivered and installed the P6 truss, which contains the first US solar arrays. The P6 was temporarily installed on top of the Z1 truss. The P6 provides solar power with solar arrays and batteries, called the photovoltaic modules.

5 ISS Assembly Mission LF 1 STS-114, the space shuttle's return to flight mission, delivered supplies and equipment to the international space station. An external stowage platform was installed with the assistance of space shuttle Discovery's robotic arm

and two space walkers. Also, space-walkers restored power to a failed control moment gyroscope and installed a new one.

6 ISS Assembly Mission 6A: Space shuttle Endeavour delivered racks inside the Raffaello multi-purpose logistics module to the Destiny laboratory. Canadarm2, the station's robotic arm, walked off the shuttle to its new home on the international space station. An ultra-high frequency (UHF) antenna that provides space-to-space communications capability for US-based space walks was installed on the station.

7 ISS Assembly Mission 11A: Space shuttle Endeavour delivered the first port truss segment, P1 truss, that was attached to the central truss segment, S0 truss. Additional cooling radiators were delivered but remained stowed until ISS Assembly Mission 12A.1. A cart, known as the crew and equipment translation aid, was delivered to help space walkers move equipment along the integrated truss structure.

8 ISS Assembly Mission 1R: The Zvezda service module was launched atop a Russian proton rocket from Baikonur Cosmodrome on 12 July 2000. The Zvezda provides living quarters and performs some life-support system functions on the international space station.

9 The international space station is viewed from space shuttle Atlantis after undocking on Tuesday, 19 June 2007 at 10:42 am.

Picture captions courtesy of NASA, 2007.

Media–ICT

Location: Barcelona, Spain
Dimensions: 145 ft x 145 ft x 125 ft high
Materials: Steel, glass, and ETFE
Architects: Cloud 9 (Enric Ruiz Geli)
Engineers: BOMA, PGI Grup Engineering

This exemplary new mediatheque building in Barcelona acts as a showcase for a whole series of technological and programmatic approaches to architecture. Described by the designers as having a hybrid program, the building is a mixture of commercial and public spaces for new technology start-up companies and more established media organizations. With a nod to the industrial heritage of the site, the architects took an early decision to invest a larger percentage of the total construction budget into the steel structure of the building (40% of the total construction cost as opposed to a more typical industry average of 25-30%). The building structure is formed by two steel-framed flank-walls, which then allows the two-floor "inhabitable" truss to be pre-assembled on the ground and winched into position using the structural walls as scaffolds. The steel gantry acts as a bridge support for the six floors, which are suspended from the truss on 6 in diameter tensile steel rods, as opposed to 20 in diameter columns had these supports been acting in compression. The logic of the suspended structure also creates a clear span ground floor between the southwest and northeast walls and

to emphasize the nature of the structural intent, all internal tensile steel elements are painted with bioluminescent paint, which glows in the dark. Interestingly, the overall dimensions and proportions of Media-ICT take their inspiration from New York's Ford Foundation building by Kevin Roche John Dinkerloo Associates (1967), where office accommodation is pushed up and outwards in its cuboid form to create a huge central garden/atrium.

The building structure is laterally stiffened by a network of diagonal steel bracing tubes that crisscross the façades, with the size and position of this eccentric diagonal grid determined by a Finite Element Analysis (FEA) of structural stiffness. The steel tubes are connected and fixed back to the structure by petal-shaped plasma-cut connector plates "flowers," of which there are 120 different types.

One of the larger ambitions of this project is to create a more environmentally responsible and responsive building, which is particularly illustrated in the two Ethylene tetrafluoroethylene (ETFE) façades. The southeast-facing façade has an external skin of inflatable ETFE cushions, which act as a variable sunscreen, letting in more daylight and sunlight in the winter for solar heat gain, and becoming more opaque in the summer months to protect and shade the building's inhabitants. The ETFE skin opacity is controlled through the differential movement of offset printed layers. The

southwest façade uses a nitrogen and oil-based fog machine to control the opacity of huge vertical ETFE cushions, with the fog pumped into the top of building-height cushions, creating a huge translucent façade. This is the first time such a system has been installed in a building. Other key technical features of the building include a network of hundreds of local sensors (light, heat, and humidity) that work in conjunction with the façade, heat and lighting systems via an Arduino (microprocessor) network to regulate the internal building environment. Media-ICT's roof is additionally covered with a garden encouraging biodiversity, photovoltaic cells and a rainwater collection system. The building achieves an A rated (the most efficient) Energy Performance Certificate.

1 Structural cross-section showing the central two-story truss and the centrally suspended floors.
2 Detail of southeast façade showing printed ETFE cushions.
3 Photograph of the southeast façade showing the bracing bars, "flower" connectors and ETFE cushions.

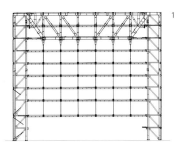

Stedelijk Museum Extension

Location: Amsterdam, The Netherlands
Dimensions: Height 59 ft, length 328 ft,
width 131 ft
Materials: Steel frame and Fiber
Reinforced Polymer composite panels
Architects: Benthem Crouwel Architects
Engineers: Arup

The building skin of this extraordinary museum addition is one of a small (but increasing) number of buildings that utilize the technology of fiber reinforced polymers (FRP) in construction. Widely used in the fabrication of yachts, aeronautical engineering and high performance sports products such as tennis rackets, this is the first time that this composite has been used for a large-scale building façade.

The new museum extension houses the largest free-span exhibition space in Amsterdam, a black box performance and audio visual room, offices, museum shop, restaurant, and library. In addition the distinctive cantilevered peak re-orientates and marks the entrance to the museum on to the Museumplein (museum square) connecting the Rijksmuseum, Van Gogh Museum, and the Concertgebouw.

The architect Mels Crouwel explains that the white coloring of the museum addition was inspired by the former Stedelijk Museum Director Willem Sandberg, who, during his tenure in 1945-1962, paired down the interior of the original 19th century building, developed a distinctive typographic identity and painted the interior white. The new addition uses a steel framed structure clad with composite panels. These 271 prefabricated elements (some as large as 49 ft x 10 ft) are attached to the precision steel internal framework with 1,100 specially designed aluminum anchor connections. After installation the separate panels are

bonded together on site to create a seamless monolithic structure. The composite "sandwich" panels consist of carbon and aramid (Twaron®) fiber laminates and special resins in order to produce a façade with a low thermal expansion coefficient, crucial for the 328 foot long construction. The composite panels were manufactured by Holland Composites and cover an area of approximately 9,840 square feet and weigh less than half of a typical curtain wall.

1 View showing the new museum extension with its 41 ft long cantilevered roof and entrance. The Museumplein is to the right of the image.
2 Cross-section drawing through the original 19th century museum and the new vessel-like addition.

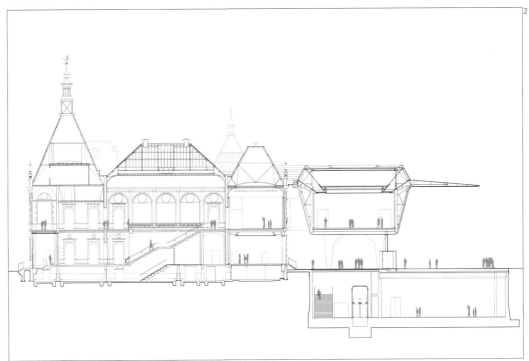

Building Codes

Health and Safety

Most of the statutory regulations that control the construction industry worldwide are there to ensure the health and safety of both construction industry operatives and end-users. Building codes cover all aspects of construction, including foundations, waterproofing, structural elements, structural stability, insulation, energy conservation, ventilation, heating, fire protection, emergency means of escape, and electric installations.

Normally architects, in association with specialist consultants such as structural and environmental engineers, must apply to the various agencies of the municipality where a project is located for a building permit prior to construction. Those agencies verify that the plans comply with applicable codes. Once construction begins, building inspectors confirm that the built work reflects the approved plans; they then issue a certificate of occupancy upon completion.

Sustainability

The world community is experiencing a growing recognition that we face imminent global crisis. Climate change threatens food and water supplies, may lead to increasingly violent storms, will endanger many species, and could raise sea levels significantly if the melting of ice caps continues at its current rate. If sea levels rise even a few feet, millions of people will be affected. Not only is the climate warming, but we also face diminishing fresh water supplies that are increasingly polluted. The hole in the ozone layer is starting to expand once more. It has become evident that oil extraction is at or near its worldwide peak, and current levels of production will not continue indefinitely.

One means of assessment that has been used to gauge the overall environmental sustainability of the world's population is the Ecological Footprint. This metric attempts to account for the productive land area necessary to provide the resources consumed, and to absorb the waste produced, by human activity. Currently, the world's population is overshooting global carrying capacity by 23 per cent. If everyone in the world lived like today's U.S. citizen, it would take 4.3 planet earths to support the world's population. Though buildings do not get as much attention as autos, they actually use more energy and create more greenhouse gas emissions that the transportation sector. Buildings account for nearly half of green house gas emissions in the USA. It is essential that we in the building industry shoulder our disproportionate share of the burden in combating the climate crisis. To do so, energy use in new buildings must be reduced dramatically to avoid the worst-case scenarios. The know-how and technology to make such dramatic cuts already exist. The challenge is in developing the will to implement changes and to educate the building community as to how and why we must do so. At the same time, we must remember that sustainability is not only about energy; we must find synergistic solutions that address a full spectrum of environmental and social issues simultaneously.

Rules Governing Construction Technology

The laws that regulate the built environment may be grouped into two broad categories: zoning and building codes. Zoning codes regulate the types of uses that may occur on a given site, site design, building massing, parking, compatibility with context, and a wide variety of other considerations. Zoning code is generally established by the city a site resides within and is tailored to the concerns of that particular community.

Building codes regulate the specific technologies implemented in any given building with the primary objectives of occupant health, safety, and fire prevention and secondary objectives of guaranteeing minimum standards of functionality and durability in the buildings. Regulating energy consumption and providing accessibility for the physically handicapped have also become major code concerns in recent years.

Codes are typically updated to reflect new knowledge and changing practices or needs of society every few years. Lessons learned from disasters such as earthquakes or major fires often lead to modifications of code requirements in the quest to guarantee the health, safety, and welfare of building occupants. The code revision process can be quite controversial because even though all stakeholders may not agree on a given provision, once incorporated into code, requirements take on the weight of law and constrain technological choice.

States or municipalities usually adopt a standard model code and modify it as required to address local conditions and preferences. The most prevalent model codes in the Unites States are the International Codes developed by the International Code Council. The most important of those codes are described here. A jurisdiction may elect to adopt only certain of the codes below and write its own rules for other sections.

Building codes rely on many standards promulgated by other organizations. For example, the ICC Electrical Code relies heavily on Standard 70 from the National Fire Protection Association (NFPA). Other commonly referenced standards organizations include ASTM International, the American National Standards Institute (ANSI), and the American Society of Heating, Refrigerating, and Air-Conditioning Engineers (ASHRAE).

International Building Code

The IBC addresses all building types and occupancies and provides regulations on building size, occupancy types, construction types, egress, fire protection, structural design, and other basic requirements. The structural sections provide span tables and standard details to simplify design of basic conditions. For more specialized conditions, an engineer may demonstrate compliance with performance standards in order to satisfy code requirements. The IBC also contains basic handicapped accessibility standards, which reflect requirements for the Americans with Disabilities Act (ADA).

International Residential Code

The IRC goes beyond the IBC in its scope of considerations to address the full spectrum of code requirements exclusively for the design of one- and two-family dwellings. Not only does it address the building structure and fire protection, but it also includes sections on plumbing, mechanical and electrical design, among others. It also has numerous prescriptive standard structural details for light frame construction.

International Fire Code

Covers fire precautions and preparedness, fire department access to buildings, fire services and systems, means of egress, and fire protection systems. Chapter 9 of the IFC addresses requirements for automatic sprinkler and other fire suppression systems in various occupancy types. NFPA 13 is referenced for specific sprinkler design requirements.

191 Rules Governing Construction Technology

International Mechanical Code

Sets out standards for heating, ventilation, and air-conditioning systems (HVAC). It requires that all spaces be ventilated by either natural or mechanical means and provides minimum fresh air volume requirements for different occupancy types. These standards are not necessarily adequate to ensure good IAQ, and should be regarded as the absolute minimum. The IMC also regulates the installation of exhaust systems, ducting, hydronic piping, refrigeration, furnaces, boilers, and other mechanical equipment types. This code specifically mandates that HVAC systems be designed and installed for energy efficiency in accordance with the International Energy Conservation Code.

International Plumbing Code

Covers both water supply and waste water systems. Water heaters, water supply and distribution, sanitary drain systems, as well as storm drainage are addressed. The IPC gives requirements for determining the numbers of fixtures of different types that are required for a given occupant load and use, and sets standards for pipe and drain sizing. The IPC approves gray water (waste water from lavatories, bathing , and clothes washing) for recycling in toilets and urinals. Gray water may be used for irrigation only if approved by local authorities.

International Fuel Gas Code

Includes standards for gas piping, equipment venting, and specific appliances such as gas fireplaces, furnaces, and clothes dryers. These appliances generally must be tested and approved according to outside standards.

International Energy Conservation Code

Establishes minimum requirements for energy efficiency in the built environment. The IECC allows for prescriptive compliance (satisfying certain code-mandated criteria on window placement, construction type, mechanical system performance, and so on for a given climate), performance-based compliance based on an energy cost budget established according to AHSHRAE 90.1, or a hybrid component performance approach for residential buildings. The IECC covers the building envelope, mechanical systems, water heating, and lighting.

ICC Electrical Code

The ICC electrical code regulates electrical systems to ensure safety and minimize the risk of fire. It references the long-established standard, NFPA 70 - the National Electric Code (NEC) for most of its technical requirements.

ICC Performance Based Code

The performance based code allows a building professional to demonstrate conformance with broad criteria for structural stability, durability, fire safety, energy efficiency, and so on by means not covered in the standardized codes. This section is used infrequently, but has the potential to allow for innovation that might not otherwise be possible under the standard codes.

192 World Commission on Environment and Development (WCED) Guidelines on Sustainable Construction

The 1987 report by the United Nations' World Commission on Environment and Development (WCED), also known as the Brundtland Report, is one of the seminal works in generating the modern conception of sustainability. It proposes at its outset what is still probably the best known definition of sustainable development: meeting "the needs of the present without compromising the ability of future generations to meet their own needs."

This charge has become a ubiquitous concern as global warming has come to be seen, almost universally, as a major threat to the future of all of earth's species. While the Brundtland definition is pleasingly simple, it does not provide much guidance on how one might actually implement sustainable development. William McDonough provided an elegant set of general principles for moving toward ecological sustainability for the 2000 World's Fair:

The Hannover Principles

1. Insist on rights of humanity and nature to co-exist in a healthy, supportive, diverse, and sustainable condition.

2. Recognize interdependence. The elements of human design interact with and depend upon the natural world, with broad and diverse implications at every scale. Expand design considerations to recognizing even distant effects.

3. Respect relationships between spirit and matter. Consider all aspects of human settlement including community, dwelling, industry, and trade in terms of existing and evolving connections between spiritual and material consciousness.

4. Accept responsibility for the consequences of design decisions upon human well-being, the viability of natural systems and their right to co-exist.

5. Create safe objects of long-term value. Do not burden future generations with requirements for maintenance or vigilant administration of potential danger due to the careless creation of products, processes, or standards.

6. Eliminate the concept of waste. Evaluate and optimize the full life-cycle of products and processes, to approach the state of natural systems, in which there is no waste.

7. Rely on natural energy flows. Human designs should, like the living world, derive their creative forces from perpetual solar income. Incorporate this energy efficiently and safely for responsible use.

8. Understand the limitations of design. No human creation lasts forever and design does not solve all problems. Those who create and plan should practice humility in the face of nature. Treat nature as a model and mentor, not as an inconvenience to be evaded or controlled.

9. Seek constant improvement by the sharing of knowledge. Encourage direct and open communication between colleagues, patrons, manufacturers and users to link long term sustainable considerations with ethical responsibility, and re-establish the integral relationship between natural processes and human activity.*

The generally accepted notion of sustainability today includes not only ecology and concern for future generations, but also the recognition that, if we are to move toward a more sustainable condition, we must achieve a greater level of social equity in the near-term and that the decisions must be economically viable to be implemented. The tension between these considerations is often depicted as a triangle with environmental impact, social equity, and economic viability at the vertices, and sustainable development occurring in the middle** A systems mindset suggests that, since the economy is a sub-system of society, and human society is but one of the many sub-systems of the environment, these considerations might be better depicted as three concentric circles with the environment represented by the outermost and the economy by the innermost.***

*From William McDonough Architects, "The Hannover Principles: Design for Sustainability," paper presented at EXPO 2000, The World's Fair, Hannover, Germany.
** From Scott Campbell, "Green Cities, Growing Cities, Just Cities?: Urban Planning and the Contradictions of Sustainable Development," *Journal of the American Planning Association*, vol. 62, no. 3, Summer 1996, pp. 296-312.
***From Roger Levett, "Sustainability indicators – integration quality of life and environmental protection," *Journal of the Royal Statistical Society*, vol. 161, part 3, 1998, pp. 291-302.

193 Leadership in Energy and Environmental Design (LEED) Green Building Rating System

In the building industry, the process of building in a way that contributes to the overall sustainability of society is often referred to as green building. In the United States, there are numerous systems that have been developed for evaluating how "green" a particular building project is. Individual cities were the first to develop such standards in local voluntary programs. For example, the city of Austin, Texas, was one of the first to develop its own green building rating system and had been very influential in the development of other such programs. Local programs are usually tailored to specific local concerns.

A number of organizations have developed green-building evaluation systems intended for national or international implementation. By far the most widely adopted of these is the Leadership in Energy and Environmental Design (LEED) Rating System from the United States Green Building Council (USGBC) (www.usgbc.org/LEED/). LEED use a checklist system that awards points based on the degree to which a project satisfies a range of criteria. There are also basic prerequisites that must be satisfied to achieve certification. LEED ratings range from Certified to Platinum based on the number of points achieved.

LEED certification is usually voluntary, but has been adopted as a requirement by some jurisdictions, especially for publicly funded projects. The system considers five basic categories – the building site, water use, energy use, material use, and indoor environment – and provides customized credits for innovations not specifically covered by other credits. Once the basic prerequisites are satisfied, one point is available for each of the criteria shown in the table. A total of 69 points are available.

To become Certified requires 26 points, 33 is considered Silver, 39 Gold, and 52 points are necessary to attain Platinum status. Critics of checklist type systems argue that this approach could actually impede progress toward sustainability by reductively focusing on specific technologies, rather than facilitating a holistic view – one that considers the building as an integrated system which regulates flows of resources and people to and from the larger environment. Other criticisms center on most evaluation systems' lack of consideration for social factors, which are also considered to be an essential part of true sustainability. The positive side of such systems is that they are easily accessible and encourage building owners to implement green building by evaluating the practices employed and certifying those achievements with a widely recognizable stamp of approval. The exponential growth of LEED is evidence of the success of this approach.

Green-building ratings are still in their infancy. In the future, these systems will make the transition from focusing on discrete technologies as most do today, toward life-cycle assessment-based models that seek to evaluate total environmental impact of a building.

Acoustics

The science of sound insulation, isolation, and sound quality control in building.

Air-Supported Membranes

Single-membrane air structures, also known as super pressure structures, are supported by internally pressurizing a sealed fabric skin.

Beam

A spanning element or bridge between two or more points, the strength of which is defined through the cross-sectional properties.

Bending

Bending is the result of material deformation under stress.

Bracing

Bracing or cross bracing is the triangulation of a structure to resist lateral loading and shear.

Building Information Modeling (BIM)

The design and construction process represented as an integrated digital database of coordinated information, which may include structure, cost, and material specifications.

Cable Net

Structures formed by a grid of cables acting under tension to form anticlastic surfaces.

Cantilever

A vertically projecting beam fixed at only one end that requires counterbalance or triangulation to resist bending.

Catenary Curve

A curved profile generated by suspending a chain or cable creates a pure-tension model formed by its own weight acting under gravity.

Climatic Envelope

The building, as defined by its capacity to protect and insulate its inhabitants from heat, cold, wind, rain, snow, etc.

Computer Numerical Control (CNC)

Relates to the use of computer controlled machine tools for fabrication processes such as cutting, milling, punching, folding, etc.

Column

Bearing elements, typically working in compression that bring load back to ground.

Computational Fluid Dynamics (CFD)

Computer simulation of flow behavior, for use in the visualization and study of wind and water movement in relation to engineering structures.

Daylight Factor

A calculation of internal light levels in building based upon the Sky Component (SC) of visible sunlight and diffuse daylight; the Externally Reflected Component (ERC) from other buildings and the Internally Reflected Component (IRC) relating to surface finishes, etc.

Evaporative Cooling

A natural cooling phenomena caused by evaporation of water by warm dry (non humid) air.

Fabrication

Fabrication is the transformation of construction materials through a physical process, such as folding, welding, casting, bending, spinning, weaving, etc.

Finite Element Analysis (FEA)

Computer simulated structural analysis technique, subdividing a constructional element through a meshing technique for physical analyses such as stress analysis.

Structural Forces

Structural forces describe the type of physical action acting upon a given structure or structural element. The main types of describable structural force are compression, tension, torsion, and shear.

Form Finding

Finding and creating new structural forms by extracting geometric information from physical models such as the soap film models of Frei Otto or the hanging chain "catenary" models of Antonio Gaudi.

Formwork

Formwork is the mold in which reinforced concrete or other plastic substrates are cast. Molds are typically timber or steel, with resin-coated plywood currently used to achieve a very smooth surface finish.

Foundations

Where columns or other bearing elements hit ground level, foundations describe how the building loads are supported, distributed and/or anchored below grade.

Fiber Reinforced Plastic (FRP)

Also known as Fiber Reinforced Polymer, this is the generic description for manmade composite structures formed from long fibers held in a resin matrix. Examples include glass fiber and carbon fiber.

Geodesic Geometry

Invented by Richard Buckminster Fuller and derived from the geometry of an icosahedron imposed onto a spherical surface. A geodesic structure distributes loads evenly across its surface and encloses the most space with the least surface.

Health and Safety
Generic terminology read in conjunction with appropriate legislation to describe the protection of human health and wellbeing during the design, construction and subsequent occupation of buildings.

Insitu Concrete
Describes concrete cast on site and not off site in a factory. Most reinforced concrete structures, including foundations, columns, beams, and slabs are cast Insitu.

Laminating
Laminating is the layering of substrates such as glass and steel to increase strength and create large spanning elements from smaller elements, such as Glue Laminated Timber (Glulam) and Laminated Glass.

Loading
Classification of different loads acting on a structure. Dead (or static) loads describe the self-weight of a structure; Live loads describe the weight of the building's occupants/furniture; Dynamic (or imposed) loads include wind and snow loading.

Masonry
A generic description for typically low tensile load-bearing materials such as stone, brick and concrete.

Monocoque Structure
Also described as stressed skin, monocoque structures use their surfaces as key structural elements working with their internal structure to resist bending. Airplane wings and contemporary car bodies are good examples, where the skin or bodywork is also the structural framework.

Parametric Design
Software-based design approach that uses relational databases to create a dynamic computer model (graphic and informational) and maintains a consistent relationship between elements as that model is changed. *See also* Building Information Modeling (BIM).

Portal Frame
A portal frame is a simple post and beam structure that has been braced or stiffened at the corners by increasing the surface area contact at the intersection of horizontal and vertical elements. Jean Prouvé used this simple idea to create some of his most notable structures.

Prefabrication
Prefabrication is the pre-making or pre-assembly of building components, typically off site in a controlled environment for improved quality.

Precast Concrete
Reinforced concrete elements produced in a controlled environment either to achieve consistent engineering characteristics or special high quality finishes. Reconstituted Stone and Terrazzo are types of precast concrete.

Reinforced Concrete
Most concrete is cast with steel reinforcing bars "rebars," steel mesh or cables and is described as reinforced concrete. The steel provides tensile strength that "mass" concrete does not have.

Shell Structures
Deriving their strength through physical form, thin reinforced concrete shell design was pioneered by engineers such as Felix Candela, who experimented with mathematically defined forms such as the hyperbolic paraboloid, and Heinz Isler, who used form finding.

Solar Geometry
The relationship between the movement of the earth in

relation to the sun creates a specific sun-path diagram for any given building location at a specific time of the year.

Space Frame
A space frame or (space grid) is a two-wayl spanning element, which typically uses small elements to create stable structural modules, which when repeated can create large spans. Space frames do not have to be flat and a geodesic structure like Buckminster Fuller's Montreal Expo dome can also be described as a space frame.

Span to Depth
An indicative ratio used to calculate the depth of a beam or truss for a given span and with a specified material.

Stack Effect
This is a natural, "passive" phenomenon, which can be used to heat, cool, and ventilate buildings. The stack effect is the movement of air through buildings based on buoyancy (heat rises) and building height.

Sun-path Diagram
The relative perceived movement of the sun in relation to the earth's orbit around the sun creates a measurable and predictable "sun-path," with altitude (alt)

used to describe solar elevation (movement in the z dimension) and azimuth to describe movement in the x,y plane.

Sustainability

Sustainability is an integrated design, engineering, and construction approach that maintains earth resources and employs low-energy solutions for the construction and life cycle energy use of buildings.

Tensegrity

Tensegrity is a highly efficient structural system, where compression members are held apart in a pure tensile matrix. Buckminster Fuller and his then student, artist Kenneth Snelson, devised this structural invention in the 1950s.

Tensile Fabric Structures

Structures formed by non-elastic fabric acting under tension to form anticlastic surfaces.

Thermal Bridging

Where thermally conductive material directly connects (or bridges) from exterior to interior environment.

Thermal Comfort

Thermal comfort is a measure of acceptable building temperature in a given location and building type. Human comfort is the more general term, which includes building temperature, air movement and humidity levels.

Thermal Insulation

Thermal Insulation acts as a barrier to heat flow; keeping buildings warm or cool depending on their location. Thermal Insulation generally uses trapped air, in a cavity, rigid foam, mineral wool, or aerogel to prevent transfer of heat or cold.

Thermal Labyrinth

A thermal "heat sink" beneath a building utilizes a labyrinth of high surface area "crenelated" concrete walls to store heat energy. Drawing air through the labyrinth into the building (based on diurnal temperature differentials) can be used for heating or cooling.

Thermal Mass

Thermal mass is the ability of a material to store heat energy. High-density materials with very few trapped air bubbles, such as brick or concrete, have a high thermal mass.

Trombe Wall

A passive heating strategy that uses the thermal mass of a wall behind a glazed façade to store and release heat energy.

Truss

A truss is a one-way spanning element. The top and bottom chords of a truss are held apart by linear members, unlike the solid flange of a typical beam, thus reducing self-weight and creating large spans. There are many truss types including bowstring, vierendeel and warren.

Wind Rose

A wind rose describes a location specific wind map, which indicates wind direction and wind speed as averaged over time to show prevailing wind characteristics.

Wind Scoop

A wind scoop might be designed to catch air and force it downwards by orientating towards the prevailing wind. A wind scoop (or cowl) might also use negative pressure to pull air up through a building for ventilation, heating, and cooling.

Further Reading and Resources

Structural Physics

Ashby, Mike, and Kara Johnson. *Materials and Design: The Art and Science of Material Selection in Product Design*. Oxford: Butterworth-Heinemann, 2002.

Baden-Powell, Charlotte. *Architects Pocket Book* (2nd edition). Oxford: Architectural Press, 2001.

Ferguson, Eugene S. *Engineering and the Mind's Eye*. Cambridge, Mass.: MIT, 1992.

Gordon, J.E. *New Science of Strong Materials: Or Why You Don't Fall Through the Floor*. Princeton: Princeton University Press, 2006.

——. *Structures, or, Why Things Don't Fall Down*. New York: Da Capo Press, 2003.

Heyman, Jacques. *Structural Analysis: A Historical Approach*. Cambridge: Cambridge University Press, 1998.

Hunt, Tony. *Structures Notebook* (2nd edition). Oxford: Architectural Press, 2003.

Larsen, Olga Popovic. *Conceptual Structural Design: Bridging the Gap Between Architects and Engineers*. London: Thomas Telford Ltd, 2003.

Rice, Peter. *An Engineer Imagines*. London: Artemis, 1993.

Structural Elements

Baden-Powell, Charlotte. *Architects Pocket Book*, 2nd edition. Oxford: Architectural Press, 2001.

Beukkers, Adriaan and Ed van Hinte. *Light-ness: The Inevitable Renaissance of Minimum Energy Structures*. Rotterdam: 010 publishers, 2005.

Bowyer, Jack (ed.). *Handbook of Building Crafts in Conservation: A Commentary on Peter Nicholson's The New Practical Builder and Workman's Companion, 1823*. London: Hutchinson, 1981.

Ching, Francis D.K. and Cassandra Adams. *Building Construction Illustrated*. New York: Wiley, 2001.

Ford, Edward R. *The Details of Modern Architecture. Volume 2: 1928 to 1988*. Cambridge, Mass.: MIT Press, 2003.

Stacey, Michael. *Component Design*. Oxford: Architectural Press, 2001.

Sulzer, Peter. Jean Prouvé, *Complete Works*, vols 1, 2, 3 and 4. Basel: Birkhauser, 1999–2008.

Weston, Richard. *Materials, Form and Architecture*. London: Laurence King Publishing, 2003.

Structural Logic

Addis, William. *Creativity and Innovation: The Structural Engineer's Contribution to Design*. Oxford: Architectural Press, 2001.

Bathurst, Bella. *The Lighthouse Stevensons: The Extraordinary Story of the Building of the Scottish Lighthouses by the Ancestors of Robert Louis Stevenson*. London: Harper Perennial, 2005.

Billington, David P. *The Art of Structural Design: A Swiss Legacy*. London: Princeton University Press, 2003.

Chilton, John. *Space Grid Structures*. Oxford: Architectural Press, 2000.

——. *Heinz Isler*, in the series *The Engineer's Contribution to Contemporary Architecture*. London: Thomas Telford Ltd, 2000.

Critchlow, Keith. *Order in Space: A Design Source Book*. New York: Viking Press, 1970.

Fuller, R. Buckminster. *Your Private Sky: R. Buckminster Fuller, the Art of Design Science*, Joachim Krausse and Claude Lichtenstein eds. Baden: Lars Muller Publishers, 1999.

Hunt, Tony. *Second Sketchbook*. Oxford: Architectural Press, 2003.

Otto, Frei and Bodo Rasch. *Finding Form: Towards an Architecture of the Minimal*. Munich: Deutscher Werkbund Bayern, 1995.

Otto, Frei (ed.). *Tensile Structures: Design Structure and Calculation of Bldgs. of Cables, Nets and Membranes.* Cambridge, Mass.: MIT Press, 1973.

Robbin, Tony. *Engineering a New Architecture.* London: Yale University Press, 1996.

Sutherland, Lyall. *Masters of Structure: Engineering Today's Innovative Buildings.* London: Laurence King Publishing, 2002.

Wells, Matthew. *30 Bridges.* London: Laurence King Publishing, 2002.

Zerning, John. *Design Guide to Anticlastic Structures in Plastic.* London: Polytechnic of Central London, 1976.

Building Performance

Bakker, Conny and Ed van Hinte. *Trespassers: Inspirations for Eco-efficient Design.* Rotterdam: 010 Publishers, 1999.

Broome, Jon. *The Green Self-build Book.* Totnes: Green Books, 2007.

Burberry, Peter. *Environment and Services*, in the series *Mitchell's Building Construction.* London: Batsford, 1977.

Charles M. Salter Associates Inc. *Acoustics: Architecture, Engineering, the Environment.* San Francisco: William Stout Publishers, 1998.

Givonni, B. *Man, Climate and Architecture*, in the series *Architectural Science.* London: Applied Science Press, 1976.

Lord, Peter and Duncan Templeton. *The Architecture of Sound: Designing Places of Assembly.* London: Architectural Press, 1986.

Nicholls, Richard. *Low.energy.design.* Oldham: Interface Publishing, 2002.

———. *Heating, Ventilation and Air Conditioning.* Oldham: Interface Publishing, 2002.

Richardson, E.G. *Acoustics for Architects.* London: Edward Arnold & Co., 1945.

Winckel, D. Fritz. *Music, Sound and Sensation.* New York: Dover Publications, 1967.

Computational Tools

Callicott, Nick. *Computer-aided Manufacture in Architecture: The Pursuit of Novelty.* Oxford: Architectural Press, 2001.

Frazer, J., *An Evolutionary Architecture,* Architectural Association, London, 1995.

Maurin, B. and R. Motro. "Concrete Shells Form-Finding with Surface Stress Density Method", *Journal of Structural Engineering,* vol. 130, issue 6 (June 2004), pp. 961–68.

Silver, Pete, Will McLean, Samantha Hardingham and Simon Veglio. *Fabrication: The Designer's Guide.* London: Architectural Press, 2006.

Case Studies

The Bartlett Technical Handbook. London: The Bartlett: Faculty of the Built Environment, University College London, 2001.

Cossons, Neil and Harry Sowden. *Ironbridge, Landscape of Industry*. London: Cassell and Co. Ltd., 1977.

De Jong, Cees and Erik Mattie. *Architectural Competitions 1792–1949*. Cologne: Taschen, 1994.

Field, Marcus. *Future Systems*. London: Phaidon, 1999.

Frampton, Kenneth. *Álvaro Siza: Complete Works*. London: Phaidon, 2006.

Fuller, Frank M. *Engineering of Pile Installations*. New York: McGraw-Hill, 1983.

Fuller, Thomas T.K. (ed.). *Buckminster Fuller: An Anthology for the New Millennium*. New York: St. Martin's, 2001.

Gorman, Michael John. *Buckminster Fuller: Designing for Mobility*. Milan: Skira, 2005.

Jodidio, Philip. *Álvaro Siza*. Cologne: Taschen, 1999.

John, Geraint, Rod Sheard and Ben Vickery. *Stadia: A Design and Development Guide*, 4th edition. Oxford: Architectural Press, 2007.

Morrison, John. *Approximate Methods of Analysis*. Bath: Buro Happold Limited, 2002.

Nicholls, Richard. *Heating, Ventilation and Air Conditioning*. Oldham: Interface Publishing, 2002.

Peters, Nils. *Jean Prouvé, 1901–1984, The Dynamics of Creation*, from the series *Basic Architecture*. Cologne: Taschen, 2006.

Roman, Antonio. *Eero Saarinen: An Architecture of Multiplicity*. New York: Princeton Architectural Press, 2003.

Sheard, Rod. *Sports Architecture*. London: Spon Press, 2001.

Spade, Rupert and Yukio Futagawa. *Eero Saarinen*. London: Thames & Hudson, 1971.

Swaan, Wim. *The Gothic Cathedral*. New York: Park Lane, 1981.

Temko, Allan. *Eero Saarinen*. New York: George Braziller, 1962.

Thompson, P., J.J.A. Tolloczko and J.N. Clarke (eds). *Stadia, Arenas and Grandstands*. London: E & FN Spon, 1998.

Toman, Rolf (ed.). *Romanesque: Architecture, Sculpture, Painting*. Cologne: Konemann, 2004.

Web Resources

www.acoustics.com
www.ballparks.com
www.bellrock.org.uk
www.bfi.org
   (Buckminster Fuller Institute)
www.buckminster.info
www.building-regs.org.uk
www.clifton-suspension-bridge.org.uk
www.communityselfbuildagency.org.uk
www.defra.gov.uk/environment/noise/
www.expo2000.de
www.greatbuildings.com
www.ironbridge.org.uk
www.oikos.com/esb/51/passivecooling.html
   (passive cooling and the use of bio-climatic charts)
www.pgacoustics.org
www.rockwool.co.uk
   (insulation)
www.seda2.org
   (Scottish Ecological Design Association)
www.segalselfbuild.co.uk
www.skyscrapernews.com/index1.php
www.spline.nl
   (Delft Spline Systems software of architectural design)
www.steelbuildinghelp.com
www.usgbc.org/LEED/
   (website of the Leadership in Energy and Environmental Design [LEED] at the United States Green Building Council [USGBC]

Pictures and diagrams are by Pete Silver and Will McLean, except for the exceptions listed below.

Laurence King Publishing and the authors wish to thank the institutions and individuals who have kindly provided photographic material and diagrams. Numbers in bold indicate page numbers, while those in brackets indicate image number. While every effort has been made to credit the present copyright holders, we apologize in advance for any unintentional omission or error, and will be pleased to insert the appropriate acknowledgment in any subsequent edition.

8 Preston Schlebusch/Getty Images.
14 Gavin Hellier/Getty Images.
17 (1) Mike Powell/Getty Images; (3) David Davies/Getty Images.
19 (2) Mike Ashby/Granta Design.
32 (2) Image courtesy of HIGOAL FRP, China.
33 (4) Malcom Fife/Getty Images; (6) Image courtesy of HOK Architects.
37 (7) Medioimages/Photodisc/Getty Images.
39 (2) Image courtesy of LaFarge Readymix; (5) Leslie Garland Picture Library/Alamy.
41 (3) Anastasia Georgouli; (4) James Madge; (6) Courtesy of Architectenbureau Micha de Haas. Photograph by Willem Franken Architectuurfotografie; (7) James Madge; (9) Image courtesy of HOK Architects.
43 (5) Peter Griffin/Alamy; (6) G.P. Bowater/Alamy.
45 (3) Anastasia Georgouli; (5) Image courtesy of Lamisell Ltd.; (6) Matthew Noble/Alamy.
47 (6) Claretta Pierantozzi; (12) Rob Cousins/Getty Images.
51 (3) Arni Katz/Alamy; (5) Reena Gogna; (9) Image courtesy of Cellbond Ltd; (10) Anthony Willis.
59 (5) James Madge; (7) Image

courtesy of Lamisell Ltd; (9, 10) Image courtesy of Lars Spuybroek, Nox Architects.
61 (4) Tim Tomlinson; (6) Image courtesy of Edward Cullinan Architects, Photographer Richard Learoyd; (7) Richard Weston; (8) Adolphe Giraudon, Alamy. (9) Anthony Willis; (10) Image courtesy of Eva Jericna Architects.
63 (3) Houston Chronicle; (4) James Madge; (5) Michael Blann/Getty Images; (9) Peter Cook/VIEW Pictures.
67 (1) Courtesy of Air Force Center, Dübendorf; (2) Arcaid/Alamy; (3) Richard Davies; (4) Courtesy of Temple Expiatori SAGRADA FAMÍLIA; (5) Allan Williams/Arcaid.
69 (2) Victoria Pearson/Getty Images; (3) Michael Hodges/Riba Library Photographs Collection; (4) Michael Snell/Alamy; (5) Images courtesy of Lamisell Ltd; (6, 7) Grant Smith/VIEW Pictures; (9, 10) Courtesy of Birds Portchmouth Russum Architects; (11, 13) © Kenneth Snelson.
72 Chart courtesy of John Farrell, XCO2 Energy.
77 (3) Tom Walker/Getty Images.
81 (3, 4) Images courtesy of LAB Architecture Studio/Atelier Ten.
83 (2) Scientifica/Getty Images.
85 (1) Hal Savas; (2) ©NASA.
89 (1) Amanda Hall/Getty Images; (4) Atlantis Verlag, Zurich; (6) Reena Gogna.
89 (8) Diagram by Manit Rostogi.
97 (1) Michael Snell/Alamy; (2) Boban Basic/Alamy.
99 (1) Allan Williams/Arcaid.
101 (2a and 2b) Diagrams courtesy of Ben Morris, Vector-Voiltec.
103 (3) Image courtesy of High Precision Devices inc; (4, 5) Courtesy of the Welsh School of Architecture, Cardiff University.
120 Drawings by Robin Base.
123 B.I.M. screengrabs courtesy of HÛT Architecture.
125 (4) Institut du Monde Arabe, Paris, France © Iain Masterton/Alamy; (5) Image courtesy of Ben Morris, Vector-Voiltec, and dRMM Architects; (6) Image © Marek Galica/Alamy.

127 (1, 6, 7) Bandstand, De la Warr Pavilion, Bexhill, Sussex, UK. Architect: Niall Mclaughin Architects/Structural Engineer: Tim Lucas, Price & Myers 3D Engineering; (2) Refurbishment of a Napoleonic Martello tower, Suffolk, UK. Architect: Piercy Conner Architects/ Structural Engineer: Tim Lucas, Price & Myers 3D Engineering; (3–5) Membrane software – Illustrations by Natalie Savva.
129 (1–6) Ecotect environmental analysis software.
131 (1) Image courtesy of Mark Taylor, Allies and Morrison; (2, 3) Ecotect environmental analysis software.
132 © 2007 Salma Samar Damluji; John Kelly/Getty Images; Image courtesy of Darren Wolf, Arup Associates; Chris Mellor/Getty Images, Petr Svarc/Getty Images; © Vincenzo Pirozzi, Rome; Paul Thompson/Getty Images; National Library of Scotland; FR Yerbury/Architectural Association Photo Library; Rob Cousins/Getty Images; Courtesy of Hammer Museum, Los Angeles; Joe Kerr/Architectural Association Photo Library; James Madge; Houston Chronicle; Michael Hodges/Riba Library Photographs Collection; Walter's Way, Lewisham, London. Image courtesy of Chris Moxey; Allan Williams/Arcaid; Ian Lambot; Courtesy of MBM Arquitectes; Greg Elms/Getty Images; Richard Davies; Michael Blann/Getty Images; Images courtesy of Lars Spuybroek, Nox/Delft Spline Systems; Images courtesy of Wilkinson Eyre/Atelier Ten; ©NASA; Image courtesy of Enric Ruiz Geli, Cloud 9; Image courtesy of Benthem Crouwel Architekten BV.
135 © 2007 Salma Samar Damluji.
137 John Kelly/Getty Images.
139 (2) Image courtesy of Darren Wolf, Arup Associates.
141 Chris Mellor/Getty Images, Petr Svarc/Getty Images.
143 (2) © Vincenzo Pirozzi, Rome.
145 (2) Paul Thompson/Getty Images.
147 (2) National Library of Scotland.
149 (2) FR Yerbury/Architectural

Association Photo Library.
151 (1) Rob Cousins/Getty Images.
153 (2, 3) Courtesy of Hammer Museum, Los Angeles.
155 (2) Joe Kerr/Architectural Association Photo Library.
157 (2) James Madge.
159 Houston Chronicle.
161 Michael Hodges/Riba Library Photographs Collection.
165 (2) Walter's Way, Lewisham, London. Image courtesy of Chris Moxey.
167 (2) Allan Williams/Arcaid.
169 Ian Lambot.
171 Courtesy of MBM Arquitectes.
173 Greg Elms/Getty Images.
175 (1) Richard Davies.
177 Michael Blann/Getty Images.
179 (1–7) Images courtesy of Lars Spuybroek, Nox/Delft Spline Systems.
180 (1–3) Images courtesy of Wilkinson Eyre/Atelier Ten.
183 (1-9) ©NASA.
185 (2) Courtesy of Enric Ruiz Geli, Cloud 9; (3) Photograph by Luis Ros, supplied courtesy of Enric Ruiz Geli, Cloud 9.
187 Image courtesy of Benthem Crouwel Architekten BV.

Acknowledgments

Lamis Bayer; Philip Cooper; Donald Dinwiddie; John-Paul Frazer; Colin Gleeson; Maria Hajitheodosi; Samantha Hardingham; Henrietta Lynch; Annalaura Palma (picture research); Hamish Muir (design); Laura Sacha; Mark Taylor; Livia Tirone; Victoria Watson; Darren Williams

Special thanks to Sam and Louie Silver for converting metric to imperial measurements. Special thanks also go to Rachael at HÛT Architecture for both written and graphic contributions to the new section on B.I.M.